D0184421

A VERY BRITISH KILLING

The Death of Baha Mousa

A. T. WILLIAMS

JONATHAN CAPE
LONDON

Published by Jonathan Cape 2012

2 4 6 8 10 9 7 5 3 1

Copyright © A. T. Williams 2012

A. T. Williams has asserted his right under the Copyright, Designs
and Patents Act 1988 to be identified as the author of this work

This book is sold subject to the condition that it shall not,
by way of trade or otherwise, be lent, resold, hired out,
or otherwise circulated without the publisher's prior
consent in any form of binding or cover other than that
in which it is published and without a similar condition,
including this condition, being imposed on the
subsequent purchaser

The author and the publishers have made every effort to trace the
holders of copyright in quotations. Any inadvertent
omissions or errors may be corrected in future editions

First published in Great Britain in 2012 by
Jonathan Cape
Random House, 20 Vauxhall Bridge Road,
London SW1V 2SA

www.vintage-books.co.uk

Addresses for companies within The Random House Group Limited can be found at:
www.randomhouse.co.uk/offices.htm

The Random House Group Limited Reg. No. 954009

A CIP catalogue record for this book is available from the British Library

ISBN 9780224096881

The Random House Group Limited supports The Forest Stewardship Council (FSC®),
the leading international forest certification organisation. Our books carrying the FSC label
are printed on FSC® certified paper. FSC is the only forest certification scheme endorsed by
the leading environmental organisations, including Greenpeace. Our paper procurement
policy can be found at www.randomhouse.co.uk/environment

Typeset in Granjon by Palimpsest Book Production Limited
Falkirk, Stirlingshire

Printed and bound in Great Britain by Clays Ltd, St Ives PLC

CONTENTS

A VERY BRITISH KILLING

Preface

VERY LITTLE IS REMEMBERED ABOUT *the Iraq War now. It has been reduced to not much more than piecemeal images: the night sky over Baghdad, lit up by flares and explosive bursts; the statue of Saddam Hussein being pulled over by ropes attached to a tank; wailing women and men splattered with blood and shell dust after a car bomb has been detonated in a busy street market; patrolling soldiers dressed in desert combat trousers and flak jackets, carrying automatic rifles at the ready as cars and pedestrians pass by in a mirage of normality; a tanned Tony Blair expressing 'no regrets' before a panel of inquirers and a brittle voice shouting 'shame' behind him; a white tent in an Oxfordshire field, covering the corpse of a scientist who had spoken freely with a journalist about Saddam's non-existent weapons of mass destruction; a man dressed in a black suit walking through a quaint English village, a hearse crawling behind him, a crowd hushed then clapping, throwing flowers on to the car; a bearded erstwhile president of Iraq having his gums and teeth inspected roughly by a US medic; a photograph of an Iraqi man standing with his wife and two young boys, posing for a family portrait to be hung on a whitewashed wall in a small anonymous house in Basra.*

This book is about the last of these. On 15 September 2003 the man in the photograph, Baha Mousa, a hotel receptionist, was killed by British Army troops in Iraq. He had been arrested the previous day and taken to a military base for questioning. For two days he and nine other civilians had their heads encased in sandbags and their wrists bound by plastic handcuffs and were kicked and punched with sustained cruelty. A succession of guards and casual army visitors took pleasure in beating the Iraqis, humiliating them and watching them suffer in the dirty concrete building where they were held. Other soldiers, officers included, did not take part but they saw what was happening and did little if anything to stop it. Some knew it was wrong. Some weren't sure. Some were too scared to intervene. But few said a word until it was far too late.

Superficially, this book is an investigation into the crimes that were committed in relation to Baha Mousa and his fellow detainees. It examines the military police inquiry which followed Baha's death and the ultimately farcical attempt to hold people criminally responsible. But there's also a deeper story told here, which the bare facts obscure. Over the past fifty years British forces have engaged in no fewer than seventeen wars around the world. In each one there have been incidents which mirror those suffered by Baha Mousa: illegal killings, torture, cruel treatment committed by British forces. And in every conflict the institutional response has been ambivalent if not duplicitous. The public condemnation of those 'bad apples' occasionally exposed as culpable has never quite hidden a culture of contempt and indifference permeating the army and government hierarchies. Protestations that Britain respects and promotes human rights and the laws of war have been accompanied by civilian and military commanders who do everything they can to look the other way or even bury cases of abuse. Systems of criminal investigation have been established to pursue those responsible, but they are not impartial, independent or effective. Official handwringing and apology have been joined by decisions and tactics preventing proper and full examination of those same crimes and others which are similar. And claims that the rule of law governs official action have been accompanied by deliberate attempts to avoid legal scrutiny. All in all the approach is very 'British': apologetic but ultimately disdainful of the suffering of people living in distant countries.

The killing of Baha Mousa holds up a mirror to this very British attitude. It reflects a light on all those foot soldiers and staff officers, civil servants, medics and lawyers, priests and politicians involved in the crime and its investigation, from court martial to public inquiry and private litigation. But it also tells us something about who we are and what is done in our name. The images of the Iraq War may warp and fade but the photograph of Baha Mousa and his family should not. It is a reminder of what being British has meant for others.

Part 1

IN IRAQ

1

I T WAS SHORTLY AFTER DAWN and already the heat was suffocating. Outside the army field hospital near Shaibah in south-eastern Iraq, fifteen kilometres south-west of Basra, a white tent had been erected on the sand a few yards apart from the main buildings. It was a police-type marquee, similar to those used to preserve crime scenes, large enough for several men to work in comfortably, if that could be possible with the temperatures already rising well above forty-five degrees centigrade. Inside the tent there was a wooden table and on the table was a man's body.

The body, solid and broad, was rudely dressed in black cotton trousers pulled down from the waist a little way over the thighs. They were sodden and split at the back. The corpse had been kept in a morgue freezer for five days and had been brought to the tent four hours earlier. It had begun to thaw and ooze fluid in the escalating temperatures inside the tent. Signs of putrefaction were becoming evident.

Next to the torso lay other items of clothing; a white vest and a green shirt. They too were ripped as though having been cut off the body and then discarded by its side. The body was tagged, a white plastic identity bracelet on the left wrist, a black tag on the right ankle. A plastic pipe projected out of the mouth. There were large circular white pads attached to the chest and an intravenous tube was still inserted into the vein in the crook of one arm, the signs of frantic but failed attempts at resuscitation.

Dr Ian Hill entered the tent sometime after dawn on 21 September 2003. He was a stranger to Shaibah and a stranger to Iraq. Five days earlier, he'd been summoned urgently from London by a Ministry of Defence official. Hill had taken the call on a train on 16 September. The official had told him that he was needed to conduct an autopsy on someone who had died in British Army custody. Although Iraqi pathologists were quite capable of doing the work, most were on strike. One of them had been threatened with death by insurgents for collaborating with the Coalition forces. Until their safety was assured, they weren't prepared to touch another cadaver for the British. But Dr Hill had accepted the commission immediately. He was an experienced pathologist, accredited by the Home Office, a senior member of the Department of Aviation Pathology, a renowned expert on air-crash victims, a recipient of the OBE, a scientific advisor on lurid TV crime dramas in his spare time, and available for commissions to conduct difficult autopsies in high-profile criminal cases.

With little preparation, Hill had been flown the 3,000 miles to Basra and taken to Shaibah army base immediately upon his arrival. When his army transport had reached the camp just after 6 am, a large number of Iraqi men and women, gathered by the gates, shouted protests about the death of the man whose body lay waiting in the tent.

Three members of the Royal Military Police (RMP) had welcomed the doctor to the camp and briefed him on how the man in the tent had died. They'd said to him that he'd been detained by a British Army unit two days before his death. He'd been hooded with sandbags, subjected to persistent beatings over those two days, and finally restrained with his head on the floor of a latrine, a knee in the back and possibly his throat constricted by the rough hem of the hoods over his head as they were tugged backwards in an attempt to control him.

Dr Hill had then been introduced to the father of the dead man, who'd confirmed that his son had been in good health before he'd been taken into custody a week before. With these preliminaries completed, Dr Hill had been given a moment to prepare himself, to put on his green gown and scrub up.

Two other persons accompanied Hill into the tent: Staff Sergeant Daren Jay, a member of the RMP's Special Investigation Branch (SIB) who would video the procedure, and a young Iraqi pathologist whom Hill would later describe as looking 'frightened'. With their assistance Dr Hill began his work. He removed the remains of clothing still covering part of the body and, talking into a microphone, described the condition and the injuries he could see.

The nose was swollen. A patch of congealed blood had formed beneath the nostrils, merging into the moustache. The left eye was surrounded by bruising. There were grazes and contusions varying in length and depth across both sides of the forehead, about the cheeks, into the eye socket, over the lids, on the ears, over the bridge of the nose and about the chin. Many of the injuries were small, insignificant when measured individually. Others were larger, more brutal.

At the neck, the doctor saw further bruising and skin damage which could never have been interpreted as trivial. There were wide patches of scraped and scuffed flesh on either side of the neck and a grazed horizontal wound beneath the chin, running over the larynx, a rough mark that suggested a string or some other cloth pulled against or across the throat.

Dr Hill then examined the torso. Bruises peppered the chest and back and ribcage in kaleidoscopic hues of blue and black, green, yellow, purple and scarlet. They merged into broader, more insistent abrasions, with one patch of bruise nine inches by six, the size of a football, traversing the body from the middle of the stomach around the left flank. There was something similar on the back, above the buttocks on the right-hand side of the body. The ribs, lower back, stomach, chest, all suffered further wide impressions of trauma rendering the whole upper body a pallet of discolouration from throat to hips, from neck to coccyx. The same pattern was repeated on all limbs. Shoulders, upper arms, forearms, hands and fingers were blemished with cuts and small contusions. Both hands were swollen too and there were lines of deep grazing around each wrist, the marks of some kind of restraining cuff pulled tight around the bone. Thighs, knees, shins, calves, all bruised.

By the time Dr Hill finished his external examination he'd recorded ninety-three separate injuries. Despite this list there was nothing, apart perhaps from the lateral mark across the throat (and that was hardly conclusive), which would explain why the body was lying on this table, in this tent. However painful they undoubtedly were, these injuries were survivable. Dr Hill had to look further, to open the body, to see for himself, the literal meaning of autopsy.

At this point in any post-mortem, as soon as the incision is made, a transformation takes place. The body becomes a cadaver and its connection with a living personality is stretched beyond natural sympathy. Every sight becomes simply *in*-human, registering only characterless organ and tissue and fat and bone. But with the parting of flesh, the weighing of organs, the extraction of fluids, there's an expectation of revelation. Some patholo-gists approach this part of the task with a sense of epiphany; there's a natural wonder in the disclosure of what can't be known from merely looking at the outside of the body.

Dr Hill was more prosaic. He recorded a broken nose, a brain slightly swollen, internal bruising about the abdomen, four broken ribs, an empty stomach, a slight amount of faeces in the large bowel, nothing in the small. He concluded that the lack of faeces would argue against claims that the dead man had been fed recently or indeed at all for some time. Nothing else of great moment was discovered, nothing obvious to explain how death had occurred.

When he emerged from the white tent Dr Hill left the young patholo-gist to talk with the family of the dead man while he reported his findings to the RMP officers. He confirmed that a sustained attack, persistent, intense, cruel and intentional, had obviously been endured. And during this prolonged violation, the evidence of which he saw registered on the exterior and interior of the body, death had occurred. The word 'torture' wasn't used, at least not by the doctor, nor 'murder'. But the mark about the throat and everything he'd seen and been told led Dr Hill to believe that the man was ultimately the victim of 'positional asphyxia'. It meant that the man, weakened by the unrelenting battering, had been forced into a posture that would have compressed his lungs to the point where he could no longer breathe, suffocating him.

Without any rest Hill was asked to see the dead man's father and explain what he'd discovered. He agreed and sat with him, bluntly describing his findings, the injuries inside and out. The words 'ligature' and 'asphyxia' and 'restraint' were used. According to the dead man's father 'strangulation' was mentioned too. This may have been a misun-derstanding, an inaccurate translation, but it remained firmly fixed in the father's mind.

On hearing the clinical details, the father became 'increasingly distressed'. Dr Hill had difficulty finishing his explanation of what he thought had happened. But to hear precisely how one's son had been killed, how he'd suffered prior to his death, to learn of the medical terms that supposedly interpreted the horrific nature of that suffering up to the very moment that his life ended, to learn about 'positional asphyxia' as though that would in some way help explain everything, could hardly have been calming.

Dr Hill left the father soon after, but his assignment wasn't finished. The RMP had another call on his services.

2

WHO WAS THE MAN WHOSE body lay in the tent in Shaibah? The dead man's father, Daoud Mousa, was one of the few who could say. And when he was given the opportunity he would calmly state that his son's name was Baha Daoud Salim Mousa. For many years there would be little time to recall much else, but, of course, there was more to know. It was still possible to compose something of the life that was to become defined by the manner of its death.

Baha had been twenty-six years old when he was killed. He'd spent all his conscious life under Saddam Hussein and his Ba'ath Party, a period of dictatorship and prolonged war: the conflict with Iran, the first Gulf War and latterly the invasion and occupation by US and British forces. That would suggest a man whose character and history had been formed in violence and oppression. And that would be true. But even in the most extraordinary and threatening conditions, people's lives still assume a mundane character, forged not through the so-called great events of nations, but in banal tragedies and insignificant happenings.

At the time Baha Mousa was first taken into custody by the British Army unit that would kill him, he was working as a receptionist at the Hotel al-Haitham in Basra. That was where he was arrested. It wasn't his normal job, if one can speak of normality in the aftermath of an invasion. He'd taken the position only a few weeks before his arrest in order to earn a little extra income and to take advantage of the air conditioning that kept the hotel beautifully cool at night. It wasn't his only occupation. He continued to operate a car sales business during the day as well as work with his father in the Basra customs office. All this activity was necessary because of his circumstances. Six months before he was killed, Baha Mousa's wife, Yasseh Samir, whom Baha was said to love deeply, had been taken ill with cancer and died. This had left him to look after his two young sons. His responsibilities didn't end there, though. The previous year his brother had also died, during a failed hospital operation, leaving two other children in his care. Although others were involved in managing the tragic consequences of these deaths Baha Mousa was the one who held the chief responsibility for the family's protection and income. The hotel offered both a chance to make some cash and, he must have thought, some relief from his family duties. He could sleep there safely and coolly.

The predilection for taking numerous jobs wasn't unusual. People in Iraq were used to searching for income in different locations and different ways. The uncertainty of existence in a country ruled by a dictator and continually on the edge of war and sanction and violence didn't allow much relaxation. Opportunities for work had to be sought and grasped. With the coming of the Coalition the precariousness of people's lives wasn't ameliorated. For some it worsened. It's hard to imagine the sense of dislocation that would have followed the invasion and overthrow of a whole regime. But still some semblance of normality persisted. Even though the war devastated many parts of cities and towns, as one would have expected from the massive scale of the bombardment and more limited ground fighting, the habits and routines of life which had developed over the years weren't wholly disrupted. The change was more insidious, particularly after the war was declared over and the 'insurgency' emerged in its wake. Then the revenge killings by Iraqis against Iraqis, the kidnappings, the thefts, the indiscriminate bombings, became prevalent. Yet this didn't supplant the daily necessity of earning money to buy what a family needed. Chances for work that came through family and friends had to be taken whenever possible.

Baha Mousa heard about the position as night receptionist in a hotel owned by a friend of his brother. He and his father went to see the owner, Ahmad Matairi, a man who would also suffer with Baha later. They would have agreed the arrangements and Baha would have started immediately. That was the way things worked.

What else would Daoud Mousa, the dead man's father, say? The pathetically thin account hardly excavated the life. But in truth what else could be said? Only general information supplemented the pathos of Baha's bare story. He and his family were Shia, which in this convoluted world of religious divide suggested he would have had no sympathy with either Saddam, al-Qaeda or those who then resisted the Coalition forces in southern Iraq. Many thousands of Shias had been slaughtered during Saddam's reign, particularly in southern Iraq. The previous ruling Sunni elite were, by common account, the faction which provided the resistance to the occupation. They were the ones who had lost the most from the overthrow of Saddam's regime. But the division of loyalties wasn't a mutually exclusive one. Individuals from all sides of the religious spectrum were offended by their country's invasion. Stories of the rape of Iraq's resources to benefit American and British interests provoked intense outrage, and not just in Iraq. This was a story circulating the world. So could anyone be sure what Baha Mousa thought about the invasion of his country or those who had invaded? His father said that Baha was happy with the British arrival. He held no allegiance to the Ba'ath Party, was even aggrieved by the regime's failure to provide decent health care for his wife and brother. What angered him or filled him with joy or motivated his actions from day to day remains locked away in the life that was ended on Monday 15 September 2003.

There was only one final clue. If accounts were to be believed, the last words Baha Mousa uttered before he died weren't a call for fanatical resistance. They were a pitiful cry as to who would now care for his two young sons.

3

WHEN HE'D FIRST ARRIVED AT Shaibah Dr Hill had been surprised with the information that there were other Iraqis to examine. Apparently Baha Mousa, the dead man, hadn't been alone. Nine men had been detained with him and subjected to similar treatment. They had survived the ordeal and were now being held at Camp Bucca, a US Army base near Umm Qasr on the Persian Gulf. The RMP wanted an independent assessment of their condition.

Dr Hill wasn't given long. The morning after the post-mortem he was taken to Camp Bucca, the Coalition forces' Theatre Internment Facility (TIF), a vast city of white-sided tents and temporary huts surrounded by high barbed-wire fences, portable searchlights and prison watchtowers. It was set in the remorselessly flat desert land outside Umm Qasr with little but oil storage tanks on the horizon to break the monotonous view. Thousands of detainees were incarcerated there in several separately fenced compounds designed to keep inmates apart and under control, although despite the precautions bloody riots would break out on various occasions during its operation. US Air Force troops armed with pepper gas and M-16 rifles patrolled the sandy boulevards between the compounds whilst the prisoners clothed in sickly yellow fatigues sweated inside the tents or watched the guards through the wire. Boredom inside and outside the fences was endemic.

The camp's name reflected a story that ran parallel to the overthrow of Saddam. Ronald Paul Bucca was an ex-Vietnam veteran of the Green Berets who happened to have been killed in the attacks on the World Trade Center in September 2001. He was one of the New York fire marshals on duty that day and he died inside the South Tower when it collapsed. The US military police decided to name the facility after him when they took over its control from the British in April 2003. It was a choice that echoed recurrent claims that the Iraq War had something to do with punishing those who had committed the 9/11 attacks. Camp Bucca was to accommodate, process and interrogate Iraqis suspected of anything from association with Saddam's regime to acts of terror on the streets of Basra and supporting the developing insurgency. It was not unrelated to the thinking which underpinned the establishment of Guantanamo Bay.

Dr Hill's task at Camp Bucca was simple: report on the condition of six men who had been transferred from the British base where Baha Mousa had died. One by one he noted their injuries. A pattern began to emerge.

The first man had considerable bruising above the waistline on his left side and his chest was extremely tender to the touch. Hill sent him for a chest X-ray.

The second appeared elderly to the doctor, although he was only forty-nine. He also had contusions across his back. He limped badly and Hill thought deep vein thrombosis was a possibility, a condition he said could have been caused by prolonged inactivity although it might have been a pre-existing problem. Hill wanted this man seen by a specialist too.

The third sported injuries very similar to Baha Mousa: extensive bodily bruising so severe that on his arrival at Camp Bucca the previous week he had been transferred immediately to hospital. Renal failure was diagnosed, probably caused by the sustained and widespread injuries. His was a life-threatening condition and one which would have long-term consequences, Hill said. If he hadn't been treated the injuries would have killed him.

The fourth suffered from head, neck and body injuries and he complained of chest pains. Dr Hill thought they were redolent of a fierce struggle.

The fifth had a cut on the forehead and bruises across the back and thigh. He had a pain in his chest and found it difficult to breathe. The doctor sent him for hospital checks.

The sixth had fewer and lesser injuries. Only some were obvious although he told the doctor of continuing pain about his ribcage. He said he was going to get medical help when he was released and that he'd been beaten persistently over a number of days. Hill called him 'somewhat truculent' and couldn't find any signs of a broken rib.

Later, Dr Hill was given photographs of three other prisoners to judge the extent of their injuries too. These showed significant bruising and damage to areas around the kidneys, a similar pattern to those men he had seen in Camp Bucca.

Despite Hill's willingness to examine the surviving detainees there was a limit to what he could do. He pleaded to the RMP officers that he was only a forensic pathologist. These men, he said, needed a clinician. Their injuries were deep, significant and required treatment. As to the psychological impact, that too was a concern and a matter requiring specialist attention.

After completing his examinations, Dr Hill left the camp and was quickly put on a plane out of Iraq. He'd been in the country for little more than thirty-six hours. Five months later, his report was finalised, although he sent a preliminary finding to the Royal Military Police early in October 2003. It spoke of clear evidence that all the men had injuries suggesting the infliction of multiple blunt impacts. The degree of harm ranged from the mild to the fatal, it said. It looked deliberate and systematic.

4

UNEASE ABOUT THE CONDITION OF the detainees had, in truth, been expressed well before Dr Hill reported his findings. As soon as the injured men were delivered to Camp Bucca in the back of an army truck on 16 September 2003, people noticed something was wrong.

A British Territorial Army intelligence officer, Lieutenant Commander Crabbe, was responsible for British admissions to the Camp.* He was on his first deployment with the armed forces even though he'd been in the Intelligence Corps since the mid-1980s. That wasn't uncommon. Reservists wouldn't normally expect to serve in war zones. The TA had long been presented by military enrolment campaigns as a sort of adult extension to the cadets or the scouts, an outward-bound club with guns. Even now when answering the rhetorical question 'what's in it for me?' on its

---

* This isn't the real name of the British intelligence officer. His identity has been protected by judicial order.

recruitment literature, the TA shows a photograph of a snowboarder rather than a man or woman in combat. But the role became slightly more demanding after 2001. The wars in Afghanistan and Iraq meant even the 'part-timers', the 'weekend warriors' as some of the regulars contemptuously named them, were called up to serve and fight.

Lt Cdr Crabbe wasn't a combatant, though. His role was interrogation. Originally recruited in the Cold War, his language skills and particularly his knowledge of Russian were valued by military intelligence. He was trained in questioning detainees. Posted to Basra in July 2003, soon after the war was declared over, he became part of the reception team at Bucca for those Iraqis captured by British forces.

On 16 September he happened to be at Basra International Airport, the British forces' HQ in southern Iraq. His superior was returning to the UK on leave and told him, before departing, that a man had died in custody and a truckload of prisoners was to be delivered to Camp Bucca later that day. Crabbe was instructed to receive the men and keep them in custody. He returned to the camp to attend to his duties, not to respond to a crisis.

Camp Bucca was designed to accommodate Iraqis detained by both US and British forces. But it was the Americans who ruled the base. They were in charge of security and general administration and they cooked the food. The British had their own small holding area within the massive compound, where their Joint Forward Interrogation Team (JFIT) was stationed. It was surrounded by its own barbed wire as though an island within a dangerous sea but it was parasitical to the US administration. Two tents housed the interns, others were used for interrogation. This, after all, was the main purpose of the facility. Information, that scarce commodity in a war where the enemy couldn't always be easily identified, was the persistent demand. Camp Bucca was one of the sites where this information could be obtained. Prisoners would arrive having already been subjected to 'tactical questioning' or TQing as the army called it. They would be handled by the front-line troops who originally arrested any Iraqis suspected of continuing the resistance. The army units or Battle Groups would keep these men at their operating bases. After a short period of questioning, purportedly to get some useful information for immediate use on the ground, detainees would then be passed on to Camp Bucca for more intense and prolonged interrogation. Lt Cdr Crabbe was the man responsible for those sent through by UK troops. He ran JFIT and his job was to get quality intelligence. He had unlimited time to do so, as much as he needed to extract the information.

When Crabbe arrived at the camp in the afternoon of 16 September a British Army lorry was waiting for him at the gateway as he had been warned. The prisoners inside were told to dismount. He watched them as they dropped from the tailgate, noticing that one was walking strangely, 'like a penguin' he said, feet splayed, shuffling about the compound. Sergeant Paul Smith, who was one of the soldiers who had brought the truck and its occupants on the hour-long journey from downtown Basra, told Crabbe the man was 'acting up', pretending to be hurt. The lieutenant commander let this pass. His primary concern was where to put these nine men. The British compound was already nearly full. He left the men to be processed by US Army personnel in accordance with camp protocol whilst he went to make arrangements to accommodate the newcomers.

It wasn't long, however, before Crabbe was called back to the reception area. A runner found him and told him to return. It was urgent, he was told. An American doctor had demanded to see him without delay. Crabbe was taken to the US medical tent where a small coterie of US Army medics was huddled around one of the new Iraqi arrivals laid on a stretcher, an intravenous drip attached to his arm. Their faces conveyed a sense of outrage before anyone spoke. Then the American doctor drew Crabbe towards the prisoner on the stretcher. He lifted a covering blanket and revealed the torso ablaze with livid bruising. Crabbe noticed myriad orange flecks amidst the purple discolouration which he thought were burst blood vessels. He called for an interpreter. When one came forward the prisoner told him a man had died and began to cry. The lieutenant commander said he was sorry. He promised to look into it.

The doctor said he feared that a broken rib might have pierced the man's liver. He'd ordered helicopter transport to take him to Shaibah field hospital where unknown to them the body of Baha Mousa was already lying in the mortuary. Then Crabbe noticed another prisoner close by, sitting on a chair, his neck encased in a white padded brace. The doctor told him there might be a broken vertebra. He too was to be flown to hospital.

Looking about the tent, Lt Cdr Crabbe caught sight of one of the British soldiers, an officer, who'd brought the detainees to the camp. Crabbe thought he looked shifty, loitering by the flap of the tent as though too embarrassed to stay and too frightened to leave. This was Lieutenant Craig Rodgers of the 1st Battalion Queen's Lancashire Regiment (or 1QLR as it was more commonly known). But there were no polite introductions. The lieutenant commander's reaction to this soldier was little short of fierce. He strode towards him confident in his superior size and superior rank. He demanded to know how the men had received their injuries. The lieutenant felt intimidated and pleaded ignorance. Crabbe was bristling with anger. He suspected, no, he *knew* that the men must've been brutally beaten at some stage and surely by the British forces who'd arrested or detained them. This officer was acting as though nothing had happened, as though he couldn't see anything wrong. Crabbe dismissed the lieutenant brusquely, told him to get out of his sight. He left and took his escort with him in the Bedford truck.

By this time Crabbe was in something of a state of shock. He rang Captain Sian Ellis-Davies, a lawyer at Division HQ who was in charge of advising on the treatment of prisoners. That was her niche. She was a qualified solicitor, but had made her career in the army. Later to become something of a pin-up prosecutor for the Army Legal Service, she was a slight blonde with a tough core. Her role in Iraq was to help any unit decide what to do with an Iraqi whom they had arrested. She would be available on her mobile, much like a duty doctor, on call, dispensing advice. The most frequently asked question was whether a particular prisoner should be detained or handed over to the Iraqi police. She didn't venture out much from Basra airport. She only visited Camp Bucca once as part of her induction when first being deployed to Iraq a few months before. After that she stayed where she was, ensconced in the relatively secure airport.

Lt Cdr Crabbe told the captain what he'd seen. He was angry. But there was nothing Ellis-Davies could do. Not at that point. Crabbe had to calm down. It was late by then and they would deal with any problems in the morning. He had no choice but to accept the advice and eventually retired for the night.

At dawn the next day, Crabbe woke up worried about his new prisoners. He went to check on them, to see how they were. They were still sleeping deeply, apart from one who sat up when he entered the tent. Without discussion the prisoner lifted his shirt displaying a broad belt of bruises across his midriff. Another detainee then awoke. He too lifted his shirt to reveal extensive bruising. It must have been a bizarre sight. If Crabbe was livid the night before he was beside himself now. He didn't wait for the others to wake up by themselves. He roused them all and had them show him any injuries. One by one they did so. Each had some visible sign of damage.

Crabbe immediately called for the US forces doctor to come back and examine all the men again. He wanted this documented and their conditions treated. He picked up the phone and began a flurry of calls, as he phrased it later. First on the list was Captain Ellis-Davies. Bruising was appearing before his eyes, he told her. The picture he painted was of a major incident simply unveiling itself in front of him. His anger was palpable. He asked for instructions, some idea of what he should do. The response wasn't quite what he'd expected. His orders were to interrogate the injured prisoners, not to placate them. He was told *not* to ask them anything about their treatment. His job was to assess their threat. They were still detainees under suspicion and were to be treated accordingly. Everything else would be left to the Royal Military Police.

5

I F THE DETAINEES HAD THOUGHT they had entered a safe haven in Camp Bucca then it was an illusion. The camp had already achieved some notoriety with the International Committee of the Red Cross, the independent guardian of the standards of treatment of prisoners in wartime. As it happened, one of the ICRC's delegates visited the camp soon after the detainees arrived. On 22 September he was at Camp Bucca and witnessed one of those frequent disruptions caused by prisoners protesting about their detention. Stones were thrown. A US forces guard in one of the watchtowers reacted and fired into the crowd. A protestor was shot in the chest, the bullet passing through the man's body and exiting the back. The ICRC delegate saw the incident. He reported that at no point did the prisoner appear to pose a threat to the life or security of the guards. Such 'brutal measures' weren't necessary, he said. There was complete disregard for life in the guard's actions. It was a disproportionate response when other methods could easily have been used. In fact, the ICRC had explicitly recommended other less violent measures on one of its previous visits to the camp earlier in the year. Those recommendations had been ignored.

Despite the ICRC's observations, witnessed first-hand, the US Army investigation found that the shooting was justified. No remedial action was taken. The shot Iraqi was declared a legitimate casualty.

That was the environment the detainees now had to endure.

6

EVEN BEFORE LT CDR CRABBE had begun to telephone whomever he could to register his fury, a report of Baha Mousa's death in custody had already been received by the Special Investigation Branch of the Royal Military Police. It had been communicated to them barely thirty minutes after Baha Mousa's death had been discovered. But the message received hadn't induced any dramatic or immediate response.

Captain Gale Nugent, head of the SIB, had taken a call at about 10.30 pm, the day *before* the rest of the detainees were delivered to Camp Bucca. She worked out of Shaibah base and Basra Palace, where 19 Mechanised Brigade of the British Army had set up its headquarters. The building sat by a lake near the Shatt-al-Arab waterway and leached obscene opulence after Saddam Hussein had personally designed its renovation. Now, it had lost most of its rich allure, having been bleached by invasion, retreat and army occupation. But it was home to the brigade which oversaw the deployment and command of 4,500 men operating in south-eastern Iraq, an area the size of Scotland. One unit, 1QLR, consisting of about 700 men, was responsible for central Basra, the toughest of the assignments.

It had been Captain Mark Moutarde who had rung Nugent. He was from Battle Group Main, the operational centre for 1QLR. The base was deep in the city of Basra, close to the al-Tamimia district, about three kilometres away from Brigade HQ. It was where the battalion housed one of its companies and its command centre under the leadership of Lieutenant Colonel Jorge Mendonça. Tactical operations were planned and launched from here. Before the invasion, the base had been one of the state's police centres. The war had damaged some of the buildings but most were still intact, though they'd been pretty much stripped of anything useful, equipment and fixtures ripped from every room in the looting phase of the conflict. Wires hung from ceilings, debris was scattered about the grounds, little remained to distinguish the place from a building site. Even when the British Army had arrived and installed themselves in the compound, there was not much improvement. The ubiquitous Portaloos, those plastic cabins of chemically treated effluence, were now dotted about the camp, accentuating the temporary nature of the occupation. But the high compound walls were still unbroken and there was space to house in safety several hundred combatant and support troops with the mass of equipment and vehicles they needed. It was also the place where any Iraqi captured on 1QLR operations, of which there were many, would be brought for initial 'processing'. It was here that Baha Mousa and the nine other detainees were brought on 14 September.

Captain Moutarde was the battalion adjutant. His job was to act as Lt Col Mendonça's personal staff officer. He was in charge of unit discipline and general management of personnel. For the most part he was a paper-pusher but still one of the officer corps, a member of the colonel's senior command team, someone who wouldn't stay behind a desk. He would accompany the CO on operations but he was also supposed to be his eyes and ears, keeping him informed of problems with the men and women of the battalion and anything else that would impact on its organisation and fighting efficiency.

On the night of 15 September, Captain Moutarde was in his office at Battle Group Main's HQ building. He was telephoned there (by someone whose name he wouldn't be able to remember later) and told that an 'incident' had occurred. One of the prisoners picked up the previous day had died. The body was in the medical centre, the Regimental Aid Post or RAP as they called it.

It must have been a shock for Moutarde. Deaths in custody weren't that common. He hurried to the RAP to take a look. It was shortly after 10 pm. When he got to the treatment room he saw a body laid on a table, its feet pointing towards the door. He didn't examine it closely. Dr Derek Keilloh, the regimental medical officer, was there. He told him the man had died of a heart attack. That wasn't good news. But then again nor was it necessarily a disaster. It was serious enough, though, to telephone his commanding officer who was out on a mission. As adjutant he had a responsibility to let him know. Lt Col Mendonça, who had already received a message from another officer at Battle Group Main (Captain Seaman, the acting operations officer) that a detainee had collapsed, couldn't break off from his operation, but he ordered Moutarde to report the matter to the SIB and Brigade HQ in Basra Palace. He would return as soon as he could, he told him. Moutarde did as he was instructed without any obvious delay by about 10.30 pm. That was the call Captain Nugent had received and recorded in the investigation log.

But Captain Moutarde didn't withdraw from the incident at that point. Despite passing on the matter to the RMP and informing Brigade headquarters, he undertook some of his own investigations – of a type. At least, by midnight at the latest, after he had called the SIB, he had compiled a report on the death for his CO. The memorandum Moutarde produced was quite detailed. It could only have come from speaking with soldiers who had been looking after the detainees and who had witnessed what had happened. Although Moutarde would later forget that he wrote this memo, its provenance has never been questioned. And it wove a complex narrative about the death the strands of which would assume great importance as the formal investigation developed. It stated that Baha Mousa was arrested early on Sunday morning, 14 September, that he was a suspect in the killing of three RMP soldiers in August 2003 in Basra, that he had been 'misbehaving' whilst in custody, that on the night of the 15th he had removed a sandbag covering his head, that he had broken out of the plasticuffs binding his wrists, that he had struggled wildly on being restrained, that he may have banged his head on the floor in the struggle, that he had slumped over when sat up by his guards, that the soldiers had checked his pulse and his breathing but found none, that Dr Keilloh had arrived to begin emergency resuscitation within minutes. Four soldiers were named in the memo as either having taken part in the struggle or having witnessed it: Corporal Payne, Private Cooper, Private Reader and Lance Corporal Redfearn.

This memo was written, it said, for the commanding officer *and* the SIB. It set out a clear timeline. After Baha Mousa had removed his plasticuffs and the sandbag covering his head at 9.30, Payne and Cooper had tried to restrain him. They'd then become aware that Mousa had stopped breathing. Reader had come in and checked his pulse and begun resuscitation. Dr Keilloh had arrived at 9.35 pm. Mousa had been stretchered back to the RAP but had been proclaimed dead at 10.05 pm.

The story that this memo told showed that Captain Moutarde had *known* quite soon after the death had been reported that a struggle had occurred immediately before the deceased's collapse. But there was no record of Moutarde ringing Captain Nugent again to tell her this. It would take her and her team several crucial days before they pieced together the information readily available to Moutarde and his CO from the outset.

7

CAPTAIN MOUTARDE WAS NOT THE only person from Battle Group Main to call Captain Gale Nugent of the SIB that night. Sergeant Charles Colley also phoned her on his mobile at about 11.00 pm. Colley, like Nugent, was a member of the RMP although he wasn't a part of the Special Investigation Branch. He commanded five RMP officers and together they advised 1QLR on policing matters. Their main job was to train Iraqi personnel. They were embedded at Battle Group Main where Mousa had died.

This kind of arrangement had become a feature of the RMP's operations within the British Army. And it meant they held an ambiguous position amongst the other soldiers. Historically they were deployed to control troop movement, directing traffic, 'getting the army to where it needed to go'. They would often accompany the troops on missions, would even fight at their sides if necessary. In Iraq, small numbers were installed with combat units, like 1QLR, to be on hand during any action, to control the public, to intervene where a civilian might have committed a crime. Their exposure to danger was the same as for the other troops, a fact that was made quite apparent in June and August 2003 when men from these RMP detachments were the direct victims of insurgency attacks. Six were brutally murdered in a horrific assault on Basra police station. Trapped by a mob, they were slaughtered in one of the largest single casualty events for the British forces of the whole occupation. The tragedy provoked intense military and media examination. Shockingly, within a couple of months three more RMP officers were killed on 23 August 2003 in Basra, a few weeks before the detention and death of Baha Mousa. Indeed, it would be these losses, quickly communicated amongst all British troops with grapevine rapidity, which in retrospect would assume a critical factor in trying to unpick the reasons for the Mousa killing.

Despite the sacrifice and shared perils, RMP personnel were also supposed to watch over the behaviour of their comrades. Every officer, wherever he or very often she was stationed, was also part of the disciplinary structure. The military police had powers of arrest of any soldier suspected of committing a crime. Although the Special Investigation Branch would undertake the examination of serious criminal allegations, operating independently from fighting units, *every* RMP officer was under a duty to respond to suspected crimes *as* a police officer. That set those stationed with battalions apart from those they slept next to, ate and suffered fear and injury with. Embedded but apart; an unsettling combination.

Sergeant Colley happened to be wandering about Battle Group Main late that night of 15 September. He was stopped by Captain Moutarde. It isn't clear exactly at what time this took place but it must have been soon after Moutarde had seen Mousa's dead body, so sometime later than 10.05 pm when the death had been officially logged. Colley claimed he was walking outside the doors of the battalion HQ building when the adjutant had called out to him. He was told a prisoner had died and, being the senior member of the RMP at Battle Group Main, was asked to help out. Colley would claim later that Moutarde had said that the body was already being prepared for shipment to Shaibah field hospital. Colley said this wasn't the correct procedure. Embracing his role as policeman, he said the body should be left alone, preserved for further investigation. But Moutarde said the SIB had told him to put the corpse in a body bag and send it on to the mortuary. Sergeant Colley wasn't happy and he said so.

Captain Moutarde wasn't about to argue. A death in custody was a police matter and even if it had been an accident or the result of natural causes, the adjutant wasn't the proper person to oversee any inquiry. Colley, being the senior RMP officer present, at least had relevant police training.

Whether Colley then acted in accordance with an order from the adjutant, or simply took it upon himself to intervene, isn't clear. But he immediately summoned one of his colleagues, Corporal Smith, and asked her to assist him. Careful then to collect and put on his red beret, that symbol of internal discipline which soldiers loathe, Colley went to the medical centre, the RAP, and spoke to Dr Keilloh. The body of

Mousa was still lying on a cot-like bed, a medical trestle, covered up to his neck with a sheet. There was a tube poking out of his mouth, Colley noticed. He asked Dr Keilloh why it had been left there and he was informed that any pathologist carrying out a post-mortem would want to know what aid had been administered. The doctor claimed he was obliged to leave everything as it was at the point when they gave up resuscitation and declared the patient dead. It was a suspected heart attack, the doctor said, and then gave Colley a brief account of what had happened.

Dr Keilloh said he'd been summoned to the detention block, what they called the Theatre Detention Facility (TDF), where Iraqi prisoners were held in Battle Group Main, at about 9.40 pm. He'd found the man, who he now knew was called Baha Mousa, lying on the floor. He wasn't breathing. The doctor had then had Mousa stretchered to the aid post so that they could use the equipment there. But at about 10.05 pm the doctor and his small team of medics had given up their resuscitation attempts and declared the man dead. That was when Captain Moutarde had been informed.

Even if Sergeant Colley didn't think there was anything particularly suspicious in this account, he still decided to secure the medical centre as a crime scene. This was odd. One doesn't normally preserve a medical bay where a man has been treated, not unless there was reason to imagine something untoward in that treatment. Nonetheless, he took the names of all the medics at the scene (there were six including Dr Keilloh) and then asked Corporal Smith not to let anyone into the small cubicle where the body lay, nor disturb the corpse. She was told to stand at the entrance, preventing anyone gaining access. Dr Keilloh protested and told Colley that he'd already been asked to prepare the body for transportation to the morgue. But Colley wasn't having that. He left Corporal Smith to ensure nothing was touched whilst he went to the detention facility where he knew Mousa had been held prior to his supposed cardiac arrest.

It was only a short walk from the aid post to the detention facility, no more than a hundred metres. The TDF was a small building, the size of a double garage, with three rooms inside. The middle room was a disused latrine. It was all very crudely constructed: flat-roofed, rough-brick walls, small windows smeared with green vehicle paint, open doorways, a couple of Portaloos against the outside front wall, a simple guardroom with little inside except dirt and dust and exposed wiring. It was right up against the boundary wall of the compound and was set apart from the main accommodation and headquarters blocks, but not by much. Iraqis captured during operations were housed here temporarily, usually for no more than fourteen hours. That was supposed to be the rule.

When he got to the TDF, Sergeant Colley met a large number of men milling about the doorways. He spotted the man in charge of the facility, Corporal Donald Payne, standing outside, waiting. He took his name and details and then those of fourteen other soldiers who were there guarding the detainees; Douglas, Rodgers, Redfearn, Reader, Cooper, Aspinall, Graham, Stirland, Allibone, Appleby, MacKenzie, Hunt, Kenny and Bentham, all of whom would assume important roles in the days, weeks, months and years ahead. Sergeant Colley wrote down the information in his pocketbook as any policeman would do. These men could be witnesses. No particular urgency accompanied these actions. An investigation would have to be conducted as with any death in custody but that was all it was, a death not a killing. And it wasn't for Colley to undertake the inquiry. The SIB had that responsibility. Although Colley would say later that if he'd suspected a crime had taken place he would have begun those investigations, he didn't think this was the case at the time. As far as he was concerned there had been no crime and therefore no crime scene to preserve, a conclusion that in retrospect sat strangely with his initial determination to secure the medical facility.

In the middle of taking down the names of the guards in his notebook Sergeant Colley rang Captain Nugent at RMP headquarters. The call was recorded by Colley as having been made at 11.00 pm and appears amidst the written details of all the soldiers he encountered at the facility. The entry sits between the roughly scrawled names of Private Stirland and Private Allibone. Why he should have made the call at that point is a mystery. Captain Nugent, of course, already knew about the death, having been phoned earlier by Captain Moutarde. But Sergeant Colley telephoned and reported that he'd secured the medical centre and had stopped the body from being removed. Nugent told him not to worry about all that. It wasn't necessary to keep the body *in situ*. He was ordered to let the corpse be put in a body bag, tagged and transferred to Shaibah mortuary as Dr Keilloh had already been instructed. According to Colley, Nugent said the SIB would deal with everything from then on. Someone would be over in the morning to carry out the formal investigation, he was informed.

This appeared to satisfy Sergeant Colley, perhaps even relieve him of further responsibility. He finished listing the names of the guards in his pocketbook and then at some point went *inside* the TDF. Whether this was after he had taken Payne's and the others' details and after he had called Nugent or *during* the call to Nugent isn't clear. But at some stage he did go inside, saw the detainees, and spoke to them. He saw nothing to concern him either as a police officer or as a simple observer. He took down the names of the Iraqi prisoners inside. He wrote their details in his notebook, but at no stage did he see anything that would suggest to him any criminal activity, any mistreatment, anything 'amiss'. If he had, he would claim later, then he would have initiated an investigation there and then. But he hadn't. No evidence was collected about the death or how it might have occurred. The site wasn't sealed off for examination. Unlike the aid post, he did nothing to preserve the location. He didn't even try. All he recorded was that the scene was cleared by 11.10 pm. Given that the detainees stayed in the TDF for the remainder of the night until 8.30 the next morning, under guard, this can only have meant that he had dispersed some of the soldiers milling about outside the building. He did nothing else of consequence. Then he went to bed.

8

As the regimental medical officer, Dr Derek Keilloh had nearly 700 men and women to care for in an environment that brought out the worst of illness, fatigue, stress, dehydration, heat stroke, as well as traumatic injury. He also had to care for those Iraqi victims of shooting who would be 'dumped on our doorstep with parts of their bodies blown off or with gunshot wounds', he would say later. But he'd only had experience of A & E work in Gosport some years previously to help prepare him. He'd never treated a gunshot wound before. At the time he served in Iraq he was approaching his twenty-eighth birthday.

Dr Keilloh's experience in Basra was shockingly intense. He was particularly affected by two incidents which occurred soon after he arrived: the killings of Captain Dai Jones and then a few weeks later three RMP members. He had met Captain Jones soon after taking up his duties. Jones had come to see him in the medical centre at Battle Group Main with a wounded soldier. Keilloh had recommended that the soldier be taken to the field hospital. Jones had decided to accompany the wounded man in the ambulance that was then targeted by the insurgents as it left the base. When Keilloh had heard the explosion he'd rushed to the scene but he'd been too late. Captain Jones was dead and his body had already been taken away by the time Keilloh arrived. The experience had marked Keilloh, he would say later.

A few days after Captain Jones' death, Keilloh had been on hand for the RMP killings as well. This time he'd arrived before the bodies had been picked up. The explosion had again happened outside Battle Group Main and Keilloh had helped pull a survivor from the blast and taken him to the Czech field hospital in Basra, the closest to the base. This too had been a horrific and affecting experience. For a junior doctor with little knowledge of and limited familiarity with catastrophic violence it must have been deeply disturbing. More so because, by his own account, Keilloh had felt isolated and unsupported from the moment he entered Iraq. He would speak later about a lack of preparation and training prior to deployment to Basra. It had been far from ideal. And when he'd arrived he'd been given little help or advice. Keilloh thought he'd been left to fend for himself.

On the evening of 15 September, at about 9.30 pm, Dr Keilloh had finished his duties for the day in the Regimental Aid Post and was walking the short distance back to the accommodation block. He heard someone shouting for a medic. He ran to the detention facility. This was where the cry had come from, so he thought.

At first, when he entered the facility, he could only see vague outlines and shapes. It was dark inside, ghostly, with only moonlight and refracted street-lamp light penetrating the gloom. But it was enough for him to see a body on the floor. One of the soldiers in the room told him the man had collapsed. Keilloh checked for breathing and heartbeat. Nothing. Nor was there a pulse.

Keilloh attempted mouth-to-mouth resuscitation. The collapsed man had vomited into his mouth, he claimed later. But he had spat out the spew and carried on regardless.

Two medics appeared with a stretcher. He hadn't ordered them to come but there they were. Someone else must have called for them. The body was laid on the stretcher and the doctor and the rest of the medics returned to the RAP, some hundred metres away.

The medical team was galvanised. They pushed through the plastic sheeting that acted as a rudimentary screen across the aid-post door and set to work. Beneath flickering strip lights, Keilloh shouted out orders to the men. Corporal Steven Winstanley was put in charge of the body's airway. He had to make sure it was unobstructed and inserted a laryngo-scope to allow a tube to be pushed down towards the lungs. Others took it in turns to push hard down on the man's chest. They all worked in cyclical bursts; compression, ventilation, adrenalin injection. Controlled frenzy. Heat and sweat.

Dr Keilloh inserted an 'intubation tube' into the man's throat, down towards the trachea. This was to allow air to pass more easily into the lungs. Three cycles of adrenalin were administered. And amidst all this frantic activity he'd had to shout out to the corridor for the throng of soldiers outside the centre (morbidly curious, or worried – he couldn't tell) to get away. They were crowding round the entrance, disturbing the concentration of the medics. It wasn't time for an audience.

The medics worked for twenty minutes, Keilloh said. Eventually he had to call a halt. The man was dead. There was no chance of revival now. Keilloh went outside. He told Sergeant Goulding, one of his team, to inform the operations room in the headquarters building. There was an RMP member outside the RAP as well. Keilloh supposedly told him or her, he doesn't remember which, to inform her superior officer, although no record was made subsequently of anyone being told as a result of this instruction.

And that was it. Keilloh left the body with all its resuscitation para-phernalia still attached. He had a 'death kit' ready, which meant a black body bag and a toe tag. It wasn't much. He would claim later that he hadn't noticed any injuries on the man he'd tried to save. He hadn't looked closely, he said. He hadn't been interested in any *cause* of death; that was for the pathologist. He and his team had treated the man as they found him. Later, he would say that all he'd noticed was a little blob of blood at the bottom of the man's nostrils. That was it. Should he have been concerned with the condition of the man whose life he'd tried to save? Would he have been expected to look? He thought not.

Dr Keilloh's work didn't quite end there, though. Sometime between declaring the man dead at 10.05 pm and midnight, two other detainees were brought to him by one of the detention facility guards. Keilloh made the detainees undress, down to their underpants. He shone a standard lamp on them. The first complained of stomach pains. An interpreter said the man was claiming that he'd been hit. But Keilloh saw nothing. He concluded that it was muscle pain and gave him an anti-inflammatory drug by injection (diclofenac). The second detainee was troubled by his back. He also claimed to have been struck by the guards. But Keilloh again saw nothing. Muscle pain once more was diagnosed, anti-inflammatory drug prescribed. Both prisoners were sent back to the detention block. Keilloh would say that he instructed the guard to bring any other detainee with a problem to him. He would also say he'd told him to bring the two detainees whom he'd treated back to him in the morning. They never reappeared. They were loaded on to the truck at 8.30 am the next day and taken to Camp Bucca with the others. That was the extent of their medical care at Battle Group Main.

9

THERE MUST HAVE BEEN SOMETHING of a hiatus at the detention facility after Baha Mousa's death and the visit of Sergeant Colley, the RMP officer at the base. All would have become quiet for a while. The uproar of the collapse and the failed resuscitation diminished and the hot night enveloped the detainees in their little prison. But they weren't left alone completely.

Captain Gareth Seeds was out on patrol with Lt Col Jorge Mendonça that evening of the 15th. He'd been asked to accompany his CO because he'd been sitting around the stifling operations room at Battle Group Main for far too long. Mendonça thought he could do with a change of scenery; he thought he needed to get out from those four walls that had confined him for so long. Often Seeds would spend twenty-four hours at a time closeted in 'ops' coordinating the battalion's missions, their 'hard' and 'soft' knocks. It would do him good to escape for a while, the CO thought.

Seeds was a career soldier. No fast-track university-funded course followed by Sandhurst for him. He'd joined up as soon as he left school, when he was sixteen, an apprentice tradesman to begin with, and then worked his way up gradually to the rank of captain by the time he was sent to Iraq. Now, aged thirty, he was the battalion's operations officer, the 'ops', for 1QLR. During 'live' operations, his duties kept him largely inside, on the radio, talking to the men on the ground, liaising with Brigade, reacting to events, recording reports, sweating in the airless ops room. His main role was to coordinate the planning of these actions. Working with a couple of signallers and a watch-keeper, he would oversee control of the 'battle space', making sure the men had the equipment they needed and support from other units if necessary.

On the night of Mousa's death he'd been close by the commanding officer, out on patrol. They had pulled over a black Mercedes. It had looked suspicious; black tinted windows, out of place. Two men had been hauled out. There had been an AK-47 and a pistol inside. Not that carrying weapons was a crime on the streets of Basra. It wasn't unusual. Guns were everywhere. People needed to protect themselves. That was understood. But the patrol had to check all the same.

As the driver and passenger were being questioned, Seeds had overheard the CO receiving a radioed report that one of the detainees back at base had died. Captain Seaman, he believed, the officer who had taken over Seeds' role in ops for the night, had been the one who had called. Seeds had heard Lt Col Mendonça order the matter to be reported at once to the RMP and Brigade HQ. They would need to know.

The patrol had finished dealing with the Mercedes driver and passenger. Mendonça's official commander's diary records that they had escorted these men to the local Iraqi police station 'for processing', although Mendonça would say later that they had taken the 'sheikh' found in the car home. Whatever the truth of this, the patrol had been cut short soon after and the unit returned to Battle Group Main.

When the company arrived back at base, Captain Seeds went to see Captain Seaman. As the position of ops had been handed over only temporarily, Seeds needed to take back the reins. He had to receive a debrief. Of course, there was only one topic worth mentioning. Seeds was told that Brigade and the SIB had been informed about the death in custody. The detainees were still over in the detention building. That was it, as though once the matter had been communicated to the RMP there was nothing else to do or say.

Seeds decided to go to bed. But before he did, he would say later, he went to use one of the Portaloos that stood up against the front wall of the detention facility. They were the closest toilets to the accommodation block, which was the building directly opposite. He thought he might also check that all was now quiet there. It wasn't his job but still . . . maybe something had worried him, given him some cause for concern, he couldn't or wouldn't say.

But as he approached the detention block something felt wrong to him. There were no guards standing outside. That might well have been the result of Sergeant Colley clearing the scene a little while before, but Seeds wouldn't have known about that. He went inside.

There were two entrances into the building, both without doors and each leading into one of the two main rooms. Seeds chose the left-hand one. It was dark inside, only a dull light from the street lamps penetrated. There was a rank smell of sweat and urine. It was acrid. The enduring heat of the night just made it worse.

Two prisoners sat together, leaning up against the wall: an old man looking dishevelled and worn, and a young lad, equally bedraggled. Seeds called the older one 'Grandad'; this was the nickname the guards had given him. It was the only name they used. Both men were handcuffed with plastic straps. A soldier was watching over them.

Seeds was incensed.

He turned on the guard.

'What the fucking hell is going on here?' he shouted. 'Why are they cuffed like that?'

The guard told him they might escape. This was not a clever response. Captain Seeds ridiculed the soldier. Did he really fucking well think that this old man and young lad would tear off their plastic handcuffs, overpower him, overpower any other fucking guards about the detention facility and make their way out of the heavily defended and armed base? Did he?

The soldier thought not.

Too fucking right.

Seeds decided to look in the other room. To do so he had to pass the middle chamber, the old latrine. It had no door. He glanced inside. There was someone lying on the floor, covered with a sheet of cardboard, feet sticking out. Seeds thought the man was dead. But he carried on, angry, into the second main room, the one on the right of the building. Here there were several more prisoners lying about the hot and dirty floor. They looked 'exhausted'. One was in a foetal position on the floor, knees curled up into his body, writhing with pain, unable to lie still. All of them had plasticuffs on.

With his temper barely under control, Captain Seeds asked one of the guards in the room whether the RMO, the doctor, had been to see the prisoners. He was told he had, although it wasn't clear which men had been examined. Whoever it was, it didn't look as though their conditions had been treated. And what about the man in the latrine? He was sleeping, the guards said. He wasn't dead.

Captain Seeds checked the water lying around the room in plastic bottles. It was room temperature, hot, tepid at best. He sent a guard to fetch cold water.

A deepening sense of disgust gripped the captain. He ordered the guards to cut off the plasticuffs, but no one had anything that would do the job. Seeds strode off, back to the ops room in the headquarters building where he'd left his gear after returning from the patrol earlier that night. He picked up some clippers from his bag. Before hurrying back to the detention block, though, he decided to look in through the door of the battalion second in command. Major Christopher Süss-Francksen was Seeds' immediate superior. His room was right next to ops. He was still awake.

'You've got to see this,' Seeds said, and asked him to come to the detention facility.

When the two officers walked into the detention block Captain Seeds immediately began to cut off the plasticuffs from each of the prisoners. He gave them cold water too. One asked to go to the toilet. He couldn't raise himself off the floor. Seeds tried to help him up but the man was too heavy. Seeds looked around for help. But Süss-Francksen was gone. (The major had left. He wouldn't return. In fact, by his own account he never came back. For quite some time a form of selective amnesia set in and he wouldn't be able to recollect *any* visit to the detention block either that night or at any other time. Later, that would change slightly. A limited memory of the visit would come back but not much else. Even then he couldn't remember seeing the detainees in any poor 'state'. Cuffed; yes. In pain, injured; no. He'd seen no reason to stay, he said later.)

The guards were gone too. Seeds stepped outside looking for some assistance and saw Major Peter Quegan, a Territorial Army officer on his first overseas deployment, making his way to use the Portaloos. Quegan was a solicitor by profession. He was one of the many Territorial Army soldiers called up to serve in Iraq, to make up the complement of men of 1QLR. His role was to act as liaison with civilians, helping out with reconstruction projects, improving the relationship between military occupiers and Basra citizens. He wasn't a fighter. But he was quite a character, instantly recognisable with his huge Colonel Blimp-like moustache and large glasses.

Seeds asked Quegan for help, and he readily agreed. Quegan walked in to find the prisoners there in a disgusting state. He was appalled, he would say later, and tried to assist Seeds with the prisoner lying on the floor. But even between them they couldn't get the man up. Then one of the guards reappeared. Seeds must've thought he would stand more chance with the private than the major because he told Quegan he could go. Quegan didn't hang around. With the private's aid, Seeds eventually managed to take the prisoner to the toilet.

A kind of momentary calm set in after that. Captain Seeds was satisfied that the men were settled. They had been watered. He had released them from their cuffs. They may still have been in the squalid conditions of the detention block, but at least some relief had been administered.

And then Seeds went to bed. What else could he do? He'd been up nearly forty-eight hours and was exhausted. He didn't report what he'd seen. The battalion second in command had been shown what was happening and knew as much as Seeds. Despite his disgust, his outrage at the condition of the detainees under the care of his battalion, he went to sleep and had nothing further to do with the matter. The only contributions he made in the immediately ensuing investigation were to answer some questions about the circumstances of Baha Mousa's detention and to provide the Royal Military Police with an aerial photograph of Battle Group Main. That was the next day and two days later respectively. It doesn't appear as though on either occasion he mentioned what he'd witnessed in the detention block that night.

Perhaps he would've done had he realised that the detainees' ordeal did not end with his intervention. There were still eight hours left to go before they were eventually loaded on to a truck and taken away to Camp Bucca.

10

B Y THE VERY EARLY HOURS of 16 September, shortly after midnight, only a couple of hours after Baha Mousa had been killed, a whole host of people knew about the 'incident'. Of the senior officers at Battle Group Main, the second in command Major Süss-Francksen knew. The adjutant, Captain Moutarde, knew. The operations officer, Captain Seeds, knew. Captain Seaman, he who had taken on Seeds' ops role for a while, knew. The regimental medical officer, Dr Keilloh, knew. Sergeant Colley, the senior RMP officer on the base, knew. Major Quegan of the TA knew.

It was an impressive list, pretty much the complete command structure of the battalion. Of the other key personnel, the regimental sergeant major, George Briscoe, also knew as he had been with Captain Seeds and the others on patrol when the call came in that a man had died. He didn't get involved, though, going to bed as so many others seemed to have done when the squad returned to Battle Group Main.

And, of course, Lt Col Mendonça, the commanding officer, knew.

It's common wisdom that the character of any military unit is intimately connected to the character of its commander. The relationship between the two has obsessed military historians across centuries. Many have seen them as indistinguishable: the leader and his qualities defining, perhaps infecting, the force he commands. It's a familiar account: the courageous officer inspiring heroic action in his men; cowardly behaviour instilling resentment and mutiny; a slipshod, chancy character producing inefficiency and ill discipline. It's the commanding officer who sets the tone, who can instil an ethos of courage, of efficiency, of determination. Or not.

Undoubtedly, the higher the rank the more tenuous the link one would think existed between leader and troops on the ground. But at battalion level in the British Army, where a lieutenant colonel will be in charge of about 700 men, the opportunity to impose a sense of purpose and direction and even a set of values will still be significant. That's the theory.

Lt Col Mendonça attempted to imprint his character on his battalion from the beginning. When he took command of 1QLR in late 2001, the first whispers of war against Iraq were stirring. He thought that units of the army would be 'competing for a place' in the invasion. And in his opinion the battalion was simply not ready. He would say later that he 'set about transforming the battalion into a war-fighting unit', emphasising 'physical fitness, mental robustness, war-fighting spirit'. They needed to be confident of their abilities. Even at a time when the politicians denied the inevitability of war, Lt Col Mendonça was preparing his men for the fight. Large-scale manoeuvres on Salisbury Plain were instigated and according to Mendonça he outshone his fellow commanders. Not that even this period of command was without controversy. During the exercise, one of his soldiers, a seventeen-year-old recruit, attacked his squad leader with a pickaxe. Private Grant Kenyon was so incensed (he would later claim in court that he had been bullied and punched) that he attacked Lance Corporal Konrad Bisping, burying a pickaxe in his head after an argument about the exercise. The story had broken in the press and the young private was arrested on suspicion of attempted murder. Everyone was shocked by the sudden violent outburst.*

Despite this incident, Mendonça believed he'd turned the battalion around. But even with all his efforts and training successes, the battalion wasn't selected for the first wave. It was only after the real war had been declared over that 1QLR received its formal warning order of deployment to Basra. They were to take over from the Black Watch. All that preparation for pitched battle had thus come to nothing.

Still, by many accounts Lt Col Mendonça was a superb fighting commander. Time and again, officers and men would say he was an excellent leader. From his superiors he would receive total support; he was seen as a man who would deal with the most difficult circumstances

---

* Private Grant Kenyon was cleared of attempted murder but pleaded guilty to grievous bodily harm. He was sentenced to four and a half years in prison and dismissed from the army. Lance Corporal Konrad Bisping was lucky to survive, but still suffered horrendous brain injuries. He was left 95 per cent blind and with continuing damage to his mental faculties.

and deliver results on the battlefield. He was a fighting man, respected for his strict control and direction. In the view of his immediate superior, in a statement given in June 2005, he was described as having 'the necessary level of robustness to deal with any challenging situations'. None of these qualities was ever questioned before or even immediately after Baha Mousa was found dead within his compound. His commanding officer, Brigadier Bill Moore, recommended him for, and Mendonça was indeed later awarded, the Distinguished Service Order recognising 'distinguished leadership during active operations against the enemy'. Major General Lamb, the senior operational commander for British forces in Iraq, wrote an appraisal of Mendonça in November 2003 when he'd finished his tour of duty. It was written after the death of Baha Mousa and the treatment of the other detainees had come to light. The report said 'as good a performance of a regular commanding officer as I have seen on operations . . . high octane performance; impressive . . . his feel for the political dynamics was of a high order, his ability to consume work impressive . . . I have raised his potential for two-rank promotion to EXCEPTIONAL'. The British Army thought it needed a man like that in Basra towards the end of 2003.

The battalion had begun its engagement in post-war Iraq with ten days acclimatising in Kuwait. Given that the summer temperatures were persistently stuck between forty and fifty degrees centigrade, it was essential. Then they had moved into Basra.

In spite of all the subsequent news reports, the embedded journalists, the international coverage and review, it's hard to picture what Iraq was like in those twilight months after the invasion and the fall of Saddam Hussein's regime. For a while there had been a kind of dangerous uncertainty in Basra, as elsewhere in Iraq, when the official battle was won. None of the signs of 'liberation' one might have expected really became embossed in the landscape. Yes, there had been some flag waving, crowds cheering the troops, plenty of sympathetic looks at the men on the ground. And yes, those Iraqi forces who might have opposed the invaders in open warfare had melted away into the hot and dusty streets with a welcome swiftness.

But the calm wasn't pure. It was tainted by gunshots, explosions. Protesting mobs would form abruptly, as if springing from the ground like dragon's teeth. Things weren't working. Power wouldn't function.

There were shortages, of oil, food, clean water, goods of every kind. Buildings were too damaged to use safely. Roads were pitted and scarred and littered with debris. Rather than performing an intense celebration and then subsiding to get on with normal life, the people remained in a state of suspension, neither truly liberated nor truly resistant. Basra was a hostile, chaotic, dangerous, overwhelming place to be. And hot. So very hot.

First into this mix had been the Black Watch regiment. They weren't popular locally. Rough and ready. Hard. They were still in war mindset. They had been in Iraq to fight the mother of all battles and hadn't been prepared for quiet and restrained policing. When they handed over to the Queen's Lancashire Regiment, a great deal of repair had to be undertaken – or perhaps more likely complete reform of the nature of relationship between occupying force and local population. A transformation was needed; the army were there more as peacekeepers than invaders intent on inflicting defeats on the enemy in pitched battles.

None of this meant that caution was unnecessary. Real threat existed amidst the streets and avenues and back alleys. There were army casualties. The men and women in 1QLR knew some of the dead personally, had seen them around. The fact that a few of their own, three RMP officers and Captain Jones, had died immediately outside their base showed just how vulnerable they all were and how constant was the danger.

The death of Captain Dai Jones in particular had a deep impact on the troops of 1QLR. Perhaps it brought home the vulnerability and jeopardy of them all. Perhaps it presaged the nature of the fight to come; another Northern Ireland environment, an uncertain enemy, danger on every street, combat a matter of unseen sniper fire or shockingly indiscriminate bombing. The manner of Captain Jones' death was significant.

Captain Jones was twenty-nine years old when he was killed. He'd joined the army when he was seventeen, first as a private soldier in the Royal Medical Corps. Then, like Captain Seeds, he'd worked his way up the ranks. In Iraq he was serving with 1QLR as the civil-military liaison officer, a role subsequently taken over by Major Quegan. His duties included the distribution of humanitarian aid, improving local infrastructure and helping the Iraqi communities establish local councils. He was the kind of officer needed for a process of reconstruction rather than battle. His general kindliness and good humour didn't protect him. When accompanying a wounded soldier in his team to hospital, the ambulance he was travelling in was blasted by a roadside bomb. He was killed instantly.

The uniform regard for Captain Jones and the breach of convention in attacking a marked ambulance singled out this incident for the soldiers in the battalion. They must have felt that there were no rules in this place, no security in the laws of war. Everyone was at risk.

But Captain Jones' death was hardly an isolated event. In September 2003 the concentration and ubiquity of threat was intense. That month alone the soldiers of 1QLR had to contend with sniper fire, public demonstrations turning violent, rocks thrown, AK-47s and pistols and ammunition, mortars and rocket-propelled grenades, available and obvious everywhere, dead Iraqis lying on the streets in the morning shot in back of the head with their hands tied behind their backs, executed for no apparent reason, car-jacking, looting, gunfights, unexploded ordnance, complaints. It was a constant diet of danger teetering over into violence with little warning.

1QLR was poorly equipped to cope with the variety of menace and logistical demands. They had to deal with a population of 1.3 million people. And of the 650 in the battalion only about 350 were fighting troops, capable of undertaking security operations. They would have to try to stop the brutal murders on the streets, the looting, the attacks directed at Coalition forces, provide protection to those who were carrying out rebuilding works, civilian liaison, the restoration of some kind of functioning society.

Lt Col Mendonça believed he was coping. Brigade thought he was coping. They were adamant the battalion was doing a fantastic job in the circumstances, focusing on security whilst not forgetting all those other humanitarian and public-relations commitments. To make a success of juggling all these duties, Mendonça had adopted a bullish and combative role. That was his character. He was tough and uncompromising and demanded absolute commitment from his men. Even though his ambition to lead the battalion in the battle for Iraq was thwarted, his approach to the job didn't change.

When Lt Col Mendonça returned from patrol that night of 15 September he went straight to his office. Captain Moutarde came there to brief him about the prisoner's death. His report was ready. It told the story of the detainee, suspect, troublesome, potentially dangerous, restrained trying to escape. What else did the commanding officer need to know? According to Mendonça he walked the short distance from his office to the detention facility after receiving Moutarde's report. He didn't enter the block. He would say later that he didn't go in and see the detainees; he only spoke with the soldiers outside. Nor did he go to see the doctor. He wasn't interested in examining the body of the man who had died. At the time, he didn't know other prisoners had suffered injury, he would say. But then, he made no attempt to find out about their situation or condition. The whole matter was left for the SIB.

Shortly after midnight, Lt Col Mendonça went to bed.

11

*T*HE BRITISH ARMY WAS NO *stranger to Iraq. It had a history there. Men had fought and died in this land for nearly a century, a fact that members of 1QLR came to know very well. They had stumbled upon the war cemetery in Basra, housing the graves of over 2,000 men of the British Empire killed in the battles of the First World War. The graveyard was in a state of dishevelment and abandon, left to decay by Saddam's regime. Although still under the nominal care of the Commonwealth War Graves Commission, that intensely dedicated organisation which maintains the silent beauty of perfect rows of identical alabaster gravestones in close-cropped grass enclosures scattered around the world, the Basra cemetery was a dusty, weed-infested compound with gravestones cracked and broken and standing at precarious angles, likely to topple with the next sharp gust of desert wind. The battalion did what it could to tidy the cemetery and make it look a little more respectable. Fragments of fractured gravestones were collected and laid out neatly to wait for proper care by the War Graves Commission. Not that official attendants would be coming any time soon. It was too dangerous for that.*

As the men of 1QLR restored some order to the cemetery they might have wondered what the men lying in those ill-tended graves had been doing there, as perhaps many were asking the same question of themselves. And the answer for the dead would have been just as convoluted and unconvincing as the answer for the living, though they might have recognised one connecting factor: oil.

In 1914, when the land was known far more mellifluously as Mesopotamia,

British troops had been sent to the region to protect British imperial strategic interests. Soon after declaration of war against Germany an expeditionary force from the British Indian Army was transported across the Indian Ocean and landed at Abadan, in Persia, adjacent to the border with Turkish-controlled territory in what is now the tip of south-eastern Iraq.

Abadan was the location of one of the world's first great oil refineries, owned and operated by the Anglo-Persian Oil Company, antecedent of British Petroleum. The importance of oil was even then profound. It fuelled the all-powerful British Navy. The Abadan refinery had to be protected from its most obvious and immediate menace, the Ottoman Empire, which then ruled Mesopotamia as well as those vast desert lands of Arabia, Palestine and Syria. Being an ally of Germany, it was likely to come in on their side. It had modern-ised its armed forces with German military support and expertise and its armies were numerous and reasonably well equipped. But despite the close relations between the Ottoman and German empires, the Turks had no particular interest in becoming entangled in the dangerously complex web of conflict that had erupted after Germany had invaded Belgium and France and marched to war with Russia. Ottoman rule was in a state of decay and a prolonged war against the vast military resources of Russia in the east and Britain in the south was the last thing it desired. It chose to stay neutral and, for the moment at least, defy German pressure to join in their campaigns.

The neutrality did not last, however. In one of those twisting knots of inter-linked events, those minor happenings with massive unforeseeable consequences, Turkey was thrust into the conflagration against its will. It began when two German warships, the SMS Goeben *and SMS* Breslau *under the command of Rear Admiral Wilhelm Souchon, evaded pursuit by the British Mediterranean fleet in October 1914 and sought refuge in Istanbul. But this escape had legal implications for the Ottoman Empire. It was bound by treaty to prevent German warships from passing through the Dardanelles. And yet it didn't want to betray its ally and refuse safe haven for its two warships. A legal solution was sought and found; the German state presented both ships as gifts to the Turkish Navy. They were handed over wholesale with German crew and officers included. Bizarrely, Admiral Souchon was then appointed as commander-in-chief of the Ottoman Navy a month later. Whether this was a deliberate act of alliance or simply recognition of the superior skills of German seamanship wasn't clear.*

Whatever the motive for the appointment, it was a disastrous manoeuvre for the Turks, ultimately leading to the destruction of the whole Ottoman Empire.

Within five weeks of his integration into the Turkish Navy, Souchon had taken the heavily gunned Goeben and Breslau along with several Turkish vessels and raided the Russian ports of Odessa and Sebastopol. The attacks came as a complete surprise, particularly as Turkey had maintained its neutrality with diplomatic assurances and careful avoidance of any conflict with its neighbours. Now that its ships had struck with significant loss of Russian vessels and men, the Russian, French and British alliance expected a full-scale engagement in the war on the side of Germany and the Austro-Hungarian Empire.

Somewhat unconvincingly, the Ottomans claimed that the naval attack hadn't been sanctioned, that it had been initiated independently by the German officers now serving in the Turkish Navy. They had acted without orders, the Ottomans said. Regret was expressed, but it did no good. Russia declared war and the British and French did the same on 5 November 1914.

The British Expeditionary Force in Persia was prepared. There had been an expectation of confrontation. To pre-empt any attack threatening key British possessions, they launched an offensive on 6 November, the day after the declaration of war, landing troops at Fao where the Shatt-al-Arab waterway meets the Persian Gulf. From there, the largely Indian troops forced their way northwards and entered Basra on 22 November. British troops occupied the town, beat off Turkish counter-attacks and launched their campaign north towards Baghdad. Basra became the centre of supply and operations for the remainder of the war in Mesopotamia.

T. E. Lawrence had passed through Basra in 1916, on his way to fight the dirty desert campaigns that would make him notorious. He had little to say about the place. It was an unattractive posting. There was some allure in the date-palm-lined river, with the shifting fleet of dhows and little barges and ornate, crumbling villas along the banks; but the supposed home of Scheherazade's Arabian Nights tales, of Aladdin and Sinbad, failed to encourage Lawrence's poetic tendencies. It was a slightly banal place, lacking an exotic charm which might otherwise have induced affection despite the heat, mud, flies and cholera. Its description as the 'Venice of Arabia', with criss-crossing waterways, was unconvincing.

Despite Basra's limited appeal, the British didn't leave after victory. Once the war, a bloody and desultory affair, was over, Mesopotamia was declared a British protectorate as part of the post-war settlement. Britain retained a military presence and bases were established to protect the oil wells and refineries. Within two years they were at war again. This time, though, British troops had to contend with a revolt that threatened to overwhelm them. In 1920 tribespeople rose up against *foreign rule and all the iniquities, perceived or real, that accompanied occupation. It was only when Winston Churchill, then Minister for War, ordered the deployment of two RAF squadrons to the country, that the balance of forces shifted. The use of strategic bombing, 'air policing' as it was called, designed simply to undermine the resolve of the rebels by targeting towns and villages suspected of harbouring them or just supporting the revolt, proved extraordinarily effective. There was little concern as to whether civilians were hit. In fact, attacks against civilians were seen as legitimate, a means by which the hopelessness of resistance against modern weapons could be instilled. RAF commanders spoke of 'forms of frightfulness' which could be used against 'semi-civilised people' to great effect. They contemplated phosphorus bombs, gas, shrapnel. And their targets were the non-combatant, the people living in the villages suspected of supporting the rebels. They bombed their animals, strafed them as they were ploughing or harvesting, attacked their fuel stores stocked for winter. It was brutal, indiscriminate and would today be considered a war crime, an attempt to induce a sense of terror in the enemy which would force them to capitulate. But then, it was considered justified. Air power was in its infancy and its destructiveness was only just becoming appreciated. The experience in Iraq helped shape the carpet-bombing strategy in the Second World War which led to the destruction of Dresden and Hamburg and many other German cities. Sir Arthur 'Bomber' Harris was a squadron commander in the conflict.*

The shift to air power in Iraq succeeded in its objective of suppressing the rebellion and kept British casualties to 400-odd men. The rebel leaders were forced to negotiate in the face of such an implacable enemy.

Although the revolt failed it helped establish clear nationalist leanings in the country. Aims couldn't be achieved militarily, not whilst the British had air power, so political change had to be sought by other means. A sort of independence was finally granted in 1932, when the country adopted 'Iraq' as its proper name. Britain continued to exercise a military presence, retaining a couple of bases 'to protect its interests'.

Less than a decade later those interests were reignited and British troops again had to fight in Iraq. In 1940, at the beginning of the Second World War, the Iraqi leadership of the time appeared to be constructing an alliance with Germany, a move which posed a direct risk to British oil supplies. The RAF and ground troops moved in once more. It wasn't much of a conflict this time, though. There was no German military support. Baghdad fell in a matter of days and a puppet government under British administration was quickly installed. There were few British casualties.

Only the scale of Basra had really changed by the time the British Army next returned in force to occupy Iraq in 2003. Of course, that once-nascent industry of oil production now surrounded and oppressed the town with ugly storage tanks, refineries, huge soulless plants of twisting pipes and gigantic cylinders and tubes. There were few romantic views amidst the sprawl. Date palms and crab apple trees may have kept their flourish amidst the gardens and riversides, but concrete and dust and piles of rubble from bombed buildings were more prevalent. Looking from the rooftops across the urban landscape the overwhelming sense was of a dull, grey monotony, low-rise concrete buildings interrupted only occasionally by a patch of greenery. For the British troops stationed there, the place remained the hot, grubby and dangerous environment it had been nearly a century before. Loathing was a common response and it perhaps echoed Lawrence's imperialistic belief that these lands, these provinces, 'were not worth one dead Englishman'.

With a certain historical symmetry, the main military camp in 1914 and throughout the First World War was outside Basra at Shaibah. It was the same camp which the British forces occupied in 2003. It was where Baha Mousa's body was taken to wait for its post-mortem.

12

WOULD CAPTAIN GALE NUGENT, HEAD of the Special Investigations Branch in Basra in September 2003, have acted with greater urgency if she had received Captain Moutarde's written report, the one he relayed to his commanding officer the night the death was first reported? Would learning about 'restraint' being applied immediately prior to the prisoner's collapse have prompted her to act differently? Would she have activated the investigation sooner if Sergeant Colley had rung her back after seeing the other detainees and told her about their condition? Or if Captain Seeds had informed her of the state of those men after he had attended to them during the night? Given that the exact content of the conversations Captain Nugent had with Moutarde, Colley and anyone else has never been revealed, these questions remain hypothetical.

Whatever information was actually transmitted late on 15 September, it was only in the morning of the 16th that Nugent assigned one of her team to the case. Even then, a lack of anxiety was evident. Captain Nugent dispatched a relatively junior officer under her command, Staff Sergeant Sherrie Cooper, alone to see what had gone on.

Not that SSgt Cooper was inexperienced. She had been in the forces since 1985 and a member of the RMP since 1988. It was only in 1999 that she had become an investigator. She was then a member of the SIB and had specialised in fraud cases after 2002 before deploying to Iraq in June 2003. It wasn't clear what expertise she had acquired in investigating crimes of the magnitude of murder or multiple torture. Nonetheless, having received her orders from Captain Nugent, she opened a case-file diary, as was the normal procedure. This was supposed to record every step taken by the investigators, registering anything of importance that would be useful or relevant to the inquiry. SSgt Cooper's first step was recorded as attending Shaibah hospital to confirm that Baha Mousa's body was there. The hospital was next to the army base where Cooper was stationed. It was a short walk to the morgue. She had photographs of the body taken to record the visible bruises and grazing to the face and stomach.

Next, she journeyed into Basra to the Iraqi teaching hospital to find a pathologist. But they were out on strike. This set in train the arrangements for summoning Dr Ian Hill from the UK to undertake the post-mortem.

Having dealt with these formalities, SSgt Cooper drove to Battle Group Main where the death had occurred. She had no investigatory team with her. She was in sole charge of the case. And it was her understanding that she had to find out 'what had happened and who was involved'. This wasn't such a simple task.

On arriving at the base SSgt Cooper went to see Captain Moutarde and Lt Col Mendonça. They were to provide her with a briefing about the incident. Moutarde's report (compiled the evening before and containing the description of Baha Mousa as a suspect in the August RMP killings, and his attempt to remove cuffs and sandbag hood before being restrained immediately prior to his death) would have been the adjutant's and the CO's main source of information. They would also have been able to say a little about the circumstances of the detainees' arrest, how as part of a large and coordinated sweep of hotels suspected of some involvement in either criminal or insurgent activity in Basra, an operation they called 'Salerno', A Company of 1QLR had entered the Hotel al-Haitham, had found a couple of Kalashnikovs, some grenades and a pistol in a locked bathroom, had arrested nine Iraqis there and returned them to Battle Group Main for questioning. Whatever story was communicated, Cooper doesn't appear to have stayed around the base for very long. The next entry in the case diary records that she went to see the battalion's intelligence officer, Major Michael Peebles. He was in his office down the corridor from the CO.

Peebles, recently promoted to the rank of major, had various jobs in 1QLR, but all were related to 'intelligence' in one form or another. SSgt Cooper would have been sent to see him because he was the Battle Group internment review officer (BGIRO). It was a position unknown to Major Peebles prior to embarking for Iraq, he would say later. It was specifically created for the occupation. Lt Col Mendonça had had to brief him on what it entailed. The BGIRO was to oversee the procedures for the arrest, questioning and treatment of all detainees for the Battle Group. He reported directly to his CO and was supported by a provost sergeant and his team of guards. The provost sergeant was responsible for the actual handling process. Their care of internees was only supposed to be temporary. Within fourteen hours of arrest, anyone detained should have been released or sent on to the Theatre Internment Facility at Camp Bucca. That was the ground rule.

Major Peebles' position made him the most senior officer with direct control over Baha Mousa and the nine other detainees during their time at Battle Group Main. He was the right person for SSgt Cooper to see.

When Cooper entered Peebles' office she immediately sensed a problem; he appeared to be undertaking his own investigation. She saw him instructing those under his command to 'question people and collect information'. This was not helpful. The SIB had exclusive charge of any serious investigation. It had to be this way. Their inquiries could be jeopardised if some line officer began to trample on evidence and interfere with witness testimony. SSgt Cooper politely told Major Peebles to stop. According to Cooper's log entry, Peebles attempted to explain himself. She reported him as saying that 'questions were being asked at parliamentary level'. How he knew this wasn't clear. But Peebles did as instructed and agreed to cease his own inquiries.

Before leaving Battle Group Main, SSgt Cooper was shown the detention facility and the room where Baha Mousa died. Both were empty. No order was given to seal the area. Nor was there an intensive attempt to take statements from those who may have witnessed events. But then, by that time, the detainees had already been taken from Battle Group Main. They had been loaded on to a Bedford truck at 8.30 that morning and driven off towards Camp Bucca. And with them had gone a large number of soldiers responsible for guarding them. They were away for a large part of the day, away from the base, and away from the RMP.

<h1 style="text-align:center">13</h1>

IT'S SOMETHING OF A DETECTIVE's truism to say that the first few hours after the discovery of a possible crime are vital in an investigation. Evidence can be sullied, buried, destroyed. Memories can be polluted, witnesses influenced. The quicker any material and statements can be collected the better. Ideally, those involved should be isolated from each other. Avoiding an 'agreement' about what people had seen is essential. Within a close-knit community such as an army unit, this is even more necessary.

The RMP know all this. They have a handbook. It tells them what amounts to good investigative practice. When they come upon a scene of crime they are supposed to use barrier tape, identify access routes, prevent unnecessary entry, avoid contamination of the area. Of course, in a war zone such advice appears ludicrous. What works for a barracks in Aldershot can hardly apply to downtown Basra during a bloody and dangerous insurgency. Captain Nugent was to make this point later. She would bemoan the difficulties that her team faced. How could they comply with '*best practice*' when they had to contend with the threat of attack, the heat, the language barrier, the lack of good pathology facilities, unreliable military radios, primitive road conditions, sand in computers and awful living conditions? Her litany of problems would be exhaustive, if not exhausting. But not all of these interferences applied at Battle Group Main. This was a secure army base, after all.

Still, few if any of the handbook procedures were followed either on the night of the 15th or anytime on the 16th. There is no record of any witness being spoken to in any depth during SSgt Cooper's first visit to the crime scene, no photographs being taken, no particularly useful evidence collected at all. There was no discussion between Cooper and Sergeant Colley, the resident RMP officer. His notebook entries setting out what he saw and did the night Baha Mousa died was only handed over a day later, or so he would claim. The detainees had been taken away before SSgt Cooper got to Battle Group Main, so she wouldn't have been aware that anything was wrong with them. She wouldn't have been able to see the state they were in, whether they had sustained any injuries. It wasn't until later on in the afternoon of 16 September, when Lt Cdr Crabbe, the outraged officer at Camp Bucca, began to call in his first concerns about the men delivered to his charge, that there was a hint of a bigger problem.

Even though the extent and nature of alleged crimes didn't fully materialise on the 16th, it still wouldn't be until Monday 22 September, a full six days after first being aware a death had occurred, that a complete and official scene-of-crime examination of the detention facility was conducted. Then, Staff Sergeant Stanford, assisting SSgt Cooper, finally went to the TDF to secure it. He found several 'items' (in point of fact, hoods, one with what looked like bloodstains inside) and drew a sketch plan. By that stage other prisoners must have passed through the facility. According to the commander's diary, 1QLR had arrested five Iraqi suspects on the night of 21 September and returned them to Battle Group Main for tactical questioning. It was possible they were held elsewhere on the base, but no mention of this was made in the logs. Whatever use to which the TDF was put after Baha Mousa had died and the other detainees moved on, no explanation was ever advanced for the delay in conducting basic forensics. Whether anything of importance would have been found is another matter.

SSgt Cooper didn't linger at Battle Group Main on the 16th. Instead, she travelled to see the army legal team stationed at Basra International Airport, another drive out of Basra to the central administrative hub for the British forces in southern Iraq. She met Captain Sian Ellis-Davies, the legal officer who would later that evening speak to Lt Cdr Crabbe at Camp Bucca and learn of the injuries sustained by the nine detainees previously held with Mousa. Captain Ellis-Davies gave a statement to SSgt Cooper, but only to confirm that she was handing over a copy of an order distributed across the British forces operating in Basra. The order was dated 3 September 2003 and set out the procedures for all units regarding 'apprehending, handling and processing of detainees and internees'. Nothing else of note was included in her statement. Ellis-Davies had not yet received the call from Camp Bucca so had nothing much else to offer. Cooper's mission was only to see what regulations 1QLR were supposed to have followed. It was a background fact-finding exercise, setting the administrative scene. It didn't contribute to discovering what had actually happened to Baha Mousa, which was Cooper's supposed instruction. She was still in the dark about the scale of the problem before her.

SSgt Cooper's investigations then finished for the day.

1 4

B Y THE TIME SSGT COOPER returned to Shaibah on the 16th, news of the death and the injuries sustained by the other detainees had been spreading like a replicating bacterium. Captain Ellis-Davies had heard from Lt Cdr Crabbe now and knew about all that he was discovering at Camp Bucca. She had told her superior, Major Russell Clifton, who was the senior legal advisor at Brigade in Basra Palace. He had been in touch with other officers at Brigade level. They already knew about the death and now there was this emerging parallel account of detainee abuse. The matter was filtering through the upper echelons of military command in Iraq and back in the UK. Emails and phone calls, hurried meetings, whispers, shaking heads, expletives no doubt. It wasn't panic, but it wasn't far off. Everyone wanted to know what had happened.

Even though the SIB had started their inquiries and even though officers knew that the RMP should be given exclusive charge over the investigation, people couldn't help but try to make sense of what had gone on over at Battle Group Main. Perhaps they wanted to prepare themselves. Some, no doubt, were shocked about the news; what did it signify? Was the incident contained? Or was it symptomatic of more widespread behaviour? Who was responsible, not just for the death but for the whole system of detention and treatment of prisoners? How far up the chain of command did that responsibility go?

And, of course, they knew that this kind of thing would be badly received back at home. Major Ed Fenton, the chief of staff at Brigade, said in an email posted to a number of his colleagues ('chaps' as he called them), 'Make no mistake, we may consider ourselves at the "front line of a war on terror", but I guarantee UK will not see it that way, and we cannot get away with treating people in this manner.'

'Us' and 'them'; troops on the ground struggling to combat an unseen enemy. Officials and bureaucrats back at home unable to appreciate the difficulties, unable even to imagine what had to be contended with: the heat, the dust, the violence, the pressure, the sheer scope of duties and responsibilities to make something of a country which was now suddenly rudderless. And all this with a pervading feeling of opprobrium emanating from all around the world – as well as back home – at their very presence in Iraq. It was an unenviable position. Some would say impossible.

The anxieties about how things would 'look' prompted questions. The questions prompted the need for information. And the need for information prompted the asking of more questions. It was a circle of challenge and response that couldn't avoid stepping into the territory of an investigation. The danger of fouling the path of police inquiry was there but this didn't constrain the command unduly. There were reputations at stake.

The first attempt to extract some answers was made at midday on the 16th, only a little over twelve hours after Baha Mousa was pronounced dead but before the other detainees had been examined at Camp Bucca. It came from Major Ben Richards who was the operations officer at Divisional level based at Basra International Airport. Here, the various brigades operating across south-east Iraq, including 19 Mechanised Brigade in Basra of which 1QLR was a part, were coordinated. All the strategic decisions were made here. It was the place for oversight, political liaison and planning.

Major Richards' commanding officer, Maj. Gen. Graeme Lamb, the man in charge of Division and therefore all the UK land forces in the region, had been told about the death in custody that morning. Brigadier Bill Moore (commander of the brigade that included 1QLR) had called him on a secure phone line. It was sensitive information and worthy of reporting to high command. Maj. Gen. Lamb immediately authorised Moore to initiate the SIB investigation, although in truth that had already begun. It wasn't enough for Lamb. He wanted more information and some assurances. Richards, his ops, was ordered to unearth what he could. And Richards didn't appear to be too worried about stepping on the SIB's toes in the process. His immediate concern was that his commanding officer was due to have a video telephone conference with his superiors in the UK that afternoon. He needed to have a better grasp of the detail.

Major Richards sent out a request for answers. It was a plea for material so that he could assess the damage. It was Captain Burbridge, the brigade operations officer at Basra Palace, who ferreted out the answers.

He responded to Richards' email within just over four hours. He rang Captain Seeds at Battle Group Main. Burbridge knew Seeds and thought him first rate. He trusted him. And Seeds must have obliged. Where *he* obtained the answers from is unclear, but he must have spoken with those responsible for the handling of the detainees as the responses given were reasonably detailed. Burbridge was able to go back to Major Richards at Division with a little stock of information: all correct procedures had been followed; there had been no forcible restraint on arrest; the detainee who had died had been seen by a medic within four hours of his arrival and he was examined at least once more during his detention; and he had been held for thirty-six hours during which he had been hooded for twenty-three hours and forty minutes. This extremely precise timing was followed by the justification that there 'was a requirement for the hood as a part of TQ [tactical questioning] conditioning and disorientation process'. No further explanation was given, but Major Richards was asked to note that the detainee had 'attempted escape repeatedly . . . on average every ten minutes'. Again, extraordinarily precise. And finally Captain Burbridge predicted the verdict of the post-mortem: 'heart attack'.

Several officers were now becoming troubled by these details. The inflammatory element was the practice of hooding, 'bagging', as they called it, coupled with 'conditioning and disorientation'. What exactly was going on?

The paper trail developed later that evening. At 8.49 pm Major Fenton sent his worried email message to the 'chaps': Major Robinson (Brigade intelligence officer), Major Clifton (Brigade legal advisor), Captain Burbridge and Major Landon (who was Fenton's deputy chief of staff at Brigade and in charge of personnel and logistics). The answers he demanded told a great deal about the dread that had been provoked. Fenton wanted to know the protocol: who was responsible for 'the detainee'; was it the battalion commanding officer or the person conducting tactical questioning? He wanted to know whether there was a handover between guards and those carrying out the questioning; and if so were there checks made as to the fitness and well-being of detainees? How long were the Battle Groups allowed to keep these people? And he asked nervously whether 'we' are 'still . . . allowed to keep detainees handcuffed and hooded' under their Rules of Engagement, rules designed for the occupation phase of the invasion.

This last enquiry betrayed a cause for concern, again not about the death itself but the hooding and cuffing. As far as the latter was concerned, Major Fenton showed he appreciated something of the context. He wrote that he understood 'the need to maintain the "pressure" [on prisoners] in order to get a better product', although he felt that 'we are going to have to work hard to justify this in future'. But what did he mean?

There was little need to decipher the terms 'pressure', 'product' and 'justify'. 'Pressure' related to the systematic *process* of using handcuffs and hoods; the 'product' to intelligence gained from tactical questioning of detainees who had been 'processed' by Battle Group Main after their initial capture; 'justify' to the need to find some protocol, some official condoning of the treatment prisoners were receiving. Major Fenton's *understanding* suggested a sort of conditional conscience in operation; on the one hand information from prisoners was needed which couldn't be acquired merely by interviewing them. It called for interrogation. But on the other hand, there had to be some limits on what methods could be used. Where was the line drawn? Where *should* it be drawn? And why wasn't this clear already?

Major Landon replied to Fenton's email within ten minutes. His contribution, though, was limited. He thought the Americans were going to pull out of Camp Bucca soon, he said, and the British would set up their own detention facility. (It was only a rumour, as it transpired, and wholly irrelevant to the matter in hand.) Landon concluded his message by asking his own question: he wanted to know whether there was a 'bible' setting the rules for handling detainees. That at least could be answered: no, there wasn't one. And this prompted a frenetic reconstruction of orders and manuals and handbooks that would fill the void. It was time for the lawyers to start work. Major Fenton's email had pushed a stone over the edge of a slope. It gained a momentum over succeeding weeks that would intensify, gathering pace as army lawyers and staff officers and civil servants and politicians began to take notice.

15

9.00 am, Wednesday 17 September 2003. Finally, a criminal investigation was unleashed by the SIB, although 'unleashed' might be an exaggeration. Whilst the top brass were engaged in their furious examination of the world of written ordinances, the SIB began to organise itself. SSgt Sherrie Cooper was sent back to Battle Group Main. They needed statements from witnesses. Simultaneously, SSgt Daren Jay was sent to Camp Bucca. News of the allegations by Lt Cdr Crabbe had reached the SIB too. These had to be checked. The detainees held there had to be questioned about Baha Mousa's death. Jay took five other RMP officers with him. It was going to be a long job.

When Cooper arrived back at Battle Group Main she sought out the adjutant, Captain Moutarde. There shouldn't have been that much to discuss. She had already been briefed by him and Lt Col Mendonça the day before. But now he suggested that she should speak specifically to Corporal Donald Payne. At least, that's how the conversation between them was recorded in the SIB investigation log. This was odd. Captain Moutarde had already compiled his report about the death in custody on the night of the 15th. It had been labelled for distribution to his commanding officer *and* the SIB. Moutarde wrote that it would be handed over to the investigating officer, whoever that might be, on their arrival at Battle Group Main the next day. And in that report Payne was clearly identified as one of the four men with Baha Mousa at the time he stopped breathing. Corporal Payne, privates Cooper and Reader, and LCpl Redfearn; all of them had been mentioned as there or thereabouts when there had been a struggle and Mousa had collapsed. It was Payne who Captain Moutarde described as restraining Mousa.

But if this report had been given to SSgt Cooper on the 16th when she first came to the base, then why hadn't an interview with this key figure – as well as the other three soldiers mentioned – already been requested? They had to be primary witnesses.

The SIB log entry suggests Captain Moutarde's report either couldn't have been handed over on the 16th *or* hadn't been read. It stated that only following Moutarde's request did SSgt Cooper then see Corporal Payne. And Payne told her that although he had no particular dealings with the detainees, he did 'brief the guard' and had 'periodically checked' the prisoners. He also told her that he was involved in restraining 'the deceased when he attempted to escape' and then died.

These words provoked an instant reaction. SSgt Cooper explained that she could no longer speak to him. He was a material witness, perhaps even a suspect. Anything he was asked and anything he said would have to be recorded under caution to preserve the integrity of any information revealed.

Clearly, if SSgt Cooper had already been told or read of Payne's involvement in Mousa's death, she must have known not to speak to him in the first place, except, of course, in accordance with proper police practice. Cooper was experienced enough in criminal investigations to know this basic evidential requirement. That was her job.

Cooper dismissed Payne and turned her attention to other less crucial sources of information. She went to the Regimental Aid Post to take statements from Dr Keilloh and his medical crew. There were six of them in all. They told her how they had tried to save Baha Mousa's life.

Next, she met with Sergeant Colley, the senior RMP officer at Battle Group Main, and was given a copy of his notebook entries from the night of the 15th. This recorded the details of the fourteen soldiers he'd found hanging about the detention facility soon after Baha Mousa had died. All would have to be questioned. But Cooper didn't even attempt to talk to these men. Instead, she went back to the detention block to make a sketch plan. It was very rough, more to help her colleagues understand the layout than to use as some kind of scientific piece of evidence. In the same vein she went to ask Captain Seeds for an aerial photograph of the whole base. Again, a matter of framing the investigative picture, not filling in its specific subject. (There was no record of Seeds saying anything about the night of the 15th when he had entered the detention facility after Baha Mousa's death and been disgusted by the condition of the remaining detainees.)

Before winding up her inquiries for the day and heading back to SIB HQ, SSgt Cooper reported to Lt Col Mendonça, to brief him on the progress of her investigations. He had demanded this. It was his base, after all. He had the right to be kept informed, though Cooper can't have had much to say. There were so many unresolved questions, so few leads pursued.

When Cooper finally headed back to Shaibah to discuss the case with Captain Nugent and her SIB colleagues, she must have made the short but dangerous journey conscious that the investigation was veering in multiple directions. Its strands were spreading like ivy, fingering a host of names with possible stories attached. All would have to be tracked. The fact that she now knew that a struggle had preceded the detainee's death altered the case's composition entirely. It allowed for a tangle of permutations, all of which would require scrutiny. She wouldn't be able to do that on her own.

16

WHILST SSGT COOPER WAS COMING to terms with an increasingly complex investigation, promising much legwork, her colleague SSgt Daren Jay was tackling the leads offered by the detainees. He, however, wasn't hampered by insufficient manpower. Along with his troop of five RMP investigators, he'd travelled to Camp Bucca early on 17 September. After recording the 'tag' numbers for each of the nine prisoners, their official record of incarceration, he and his team began to question them.

The investigators couldn't help but see the detainees' injuries. They took photographs. Some of the prisoners told their interviewers how they had been mistreated after their arrest on the 14th. They talked about the hitting, the kicking, about the hoods, the rasping plasticuffs. Their stories were consistent. And it wasn't long before SSgt Jay discovered that the two most seriously affected prisoners were being treated separately at Basra General Hospital and the British field hospital at Shaibah. They would have to be seen later. Even so, the picture of abuse was gradually beginning to take a sharper form.

SSgt Jay's most pressing concern, however, was about the dead man and what the detainees knew about him. Who was he? Where did he live? Who was his next of kin? None of this information was yet available to the SIB. Nothing had been handed over by Battle Group Main to help them. By cross-checking notes from the detainees' interviews, Jay was able at least to find out the man's name and roughly where he had lived in Basra. It would still take some time to locate the family home but they had sufficient information, Jay believed, to be able to track it down. He reckoned it would take some 'house-to-house' inquiries but that would be OK. It wouldn't take long.

Later that day staff sergeants Cooper and Jay both met with Captain Nugent. She had called them back to reflect on where they were with the case. Warrant Officer Spence was asked to join them. He was already being lined up to take over from Cooper, who was nearing the end of her stint in Iraq. She would be leaving theatre in a matter of weeks, as would

Nugent herself and indeed Jay and Spence before too long. That was the character of personnel shift during the occupation. Officers would arrive and be gone within six, maybe three months. It didn't help with the continuity of inquiries, although it was already decided that Spence would continue with the case after leaving Iraq.

Chaired by Nugent, the investigators began to compare their discoveries. It quickly became clear that the investigation already promised to be complex, sensitive, even political. Its parameters were shifting with great rapidity. They figured that more than thirty people would have to be interviewed. None of these could be identified as an obvious culprit, though. They didn't have a 'prime suspect'. Despite the detainees' injuries, now the subject of photographic proof, and despite Corporal Payne's admission to Cooper in the morning that he had been restraining the dead man at the time of the fatal collapse, they had nothing concrete to connect a crime with a suspected perpetrator. Of course, no scene-of-crime examination had been properly conducted so there was no forensic evidence available. No post-mortem had yet been carried out either, so Dr Keilloh's diagnosis of heart attack remained the current wisdom. And no identification had been made of any soldier actually assaulting the surviving detainees. All they had was a long list of names, Iraqi and British, which continued to lengthen. Then Captain Nugent informed her team that she had only just received a message from a Major Richard West, who was the second in command of the 3rd Regiment of Royal Military Police based at Division level, over at Basra International Airport. He had told her that he'd been approached by the media ops officer at Division who said that three of his drivers may have seen something troubling. The report coming through was that they had witnessed the detainees being abused at Battle Group Main. There was no greater detail than that but still it was the first indication of direct independent witness testimony of ill-treatment that might explain all those bruises and contusions Jay had already photographed. The drivers would have to be seen as soon as possible.

No one doubted now that this would be a serious and prolonged case. They started to plan for the task, interviews they would need to carry out, the post-mortem, the collection of photographic and medical information of the injuries sustained. This wasn't simply a matter of sending out detectives to speak to people. By Brigade orders none of them could go anywhere without proper protection. They couldn't even drive from base to base without back-up. It would require a protective force from Division, possible but still a halter on their operations. Everything would be slowed down as they waited for armed guards to accompany them on all their trips around Basra. Resources were stretched as it was.

Despite the developing demands of the investigation the SIB officers agreed that contacting the next of kin now had to take priority. SSgt Jay was given the responsibility.

17

DAOUD MOUSA, THE DEAD MAN's father, knew about his son's arrest from the beginning. By chance he was at the Hotel al-Haitham when his son was taken. It was a strange, unsettling experience, but not alarming. There was no indication of particular threat or danger to his son. It was one of those misunderstandings that should have been resolved fairly quickly. Daoud Mousa was assured by a British officer that Baha would be released within a couple of hours. That was early on the Sunday morning, 14 September. And Daoud Mousa trusted the word of the officer.

Daoud's initial faith may have had something to do with his own professional background. He had been a major in the Basra police force during the Saddam Hussein regime and knew well the process of apprehending and questioning people to obtain information. He had served in the force for twenty-four years, although his career had ended abruptly when he had been 'encouraged' to retire early in 1991. Southern Iraq had not been loyal to Saddam then. After a local uprising in that year, brutally suppressed by Saddam's Ba'ath Party, Daoud had been seen as suspect. He hadn't been part of the rebellion, but neither had he fought against it. That made him unreliable. He was forty-five. He would take up his post as police officer again in 2004, but at the time of the invasion he was working at the Basra customs office. He still carried himself with that self-assurance imprinted by years of exercising authority.

Once his son had taken the job as night receptionist at the hotel, Daoud had fallen quickly into the habit of going to pick up Baha after his night shift. Around seven to eight o'clock in the morning Daoud would meet him there to take him back home.

The two men were very close, the 'best of friends' he would say. Their lives were bound tightly together after the deaths of Baha's wife and Daoud's other son. They had assumed joint responsibility for looking after the family. In those times, soon after the Coalition had invaded, it was a hectic and unnerving enterprise to attain any kind of security. And just earning money to survive wasn't the only demand. The whole environment of Basra during those first months of the occupation was intimidating. British troops may have seen the violence on the streets and been targeted themselves, but the Iraqi citizens of Basra were hardly immune from random attack. And they possessed little means to defend themselves. They didn't have body armour or armoured vehicles. Their vulnerability was obvious and persistent. The bodies dumped regularly in backstreets, executed by one violent group or another, were proof of that.

On that Sunday 14 September, Daoud arrived at the hotel expecting to meet his son as usual. There was a host of military vehicles encircling the building. British Army guards stood at the entrance. He looked through one of the picture windows of the hotel to see what was going on. He could make out a number of soldiers standing around a safe, using some kind of hammer or crowbar to force it open. He watched as the soldiers began to pull cash from the safe and stuff it into their trousers. One of them shoved a bundle of money into his shirt. Others were filling their pockets. He could hardly believe what he was witnessing. Whether offended because of his previous profession or simply at what appeared so obviously wrong, he decided to report the matter. As he could speak a little English, he approached the guard by the hotel entrance and told him that he wanted to speak to the officer in charge, he wanted to report a crime.

Daoud had no idea what was happening in the hotel. But he was allowed inside, perhaps surprisingly given that this was the scene of a military operation, and directed to an officer referred to as 'Lieutenant Mike'. (Years later, this officer would be identified as Lieutenant Michael Crosbie. He was Territorial Army, assigned to intelligence in 1QLR when sent out to Iraq in June 2003. Daoud Mousa was not told the lieutenant's full name. He only had his description: 'chubby, white, blond . . . of average height'. It was enough to identify Crosbie.) The lieutenant asked Daoud to pick out the culprits, but Daoud could only recognise the soldier who had hidden money in his shirt. His was the only face he had really observed. By chance the soldier was standing in the hotel lobby. Daoud pointed him out to Lieutenant Crosbie. The soldier was ordered over, his pockets and clothing were checked and packets of notes emerged. There was an immediate reprimand. Daoud said the officer, clearly disgusted, shoved the soldier and told him to get out of the hotel.

The officer asked Daoud to write out a statement. He was given a red pen and scribbled his account in Arabic on some rough pieces of paper. Lieutenant Crosbie would say later that an interpreter was called to translate the note. It was handed over and was never seen again.

While Daoud was reporting the theft and writing the note, he noticed that thē hotel employees were lying face down on the floor, rifles pointed at their heads by several guards. His son was among them. When he'd finished his statement Daoud sought out Lieutenant Crosbie. He asked for his son's release. It was only fair given the information about the stealing that he'd given. The officer assured him his son would be set free within a couple of hours and wrote a telephone number where the officer could be contacted. This was enough for Daoud. He went outside and looked on as the hotel employees were brought out one by one through the entrance and loaded on to Bedford trucks. Their hands were tied with plasticuffs. Daoud was able to shout to his son not to worry, he would be released in an hour or two. These were the last words he would say to him.

Daoud Mousa went home faithfully holding on to the assurances he had received. It was disturbing to have his son arrested, but in a peculiar way he was comforted by knowing that Baha was safe. He was in the custody of the British. There was nothing to be afraid of there. Daoud could talk to their officers, he had a phone number, he could assume that whatever else might happen, his son would be looked after. Even in that turmoil of newly occupied Iraq, where the present and future were so brittle, there was reason to trust the British. They were liberators, not a conquering horde. What was there to be feared?

After a couple of hours Daoud went to the gates of Camp Stephen, the British Army base where he thought his son might have been taken. As with most of the British centres of operation this was closely associated with the previous regime. Unhappily, it was one of the houses of 'Chemical Ali', the 'King of Spades' in the USA's macabre pack of cards of wanted men issued at the beginning of the invasion. He was a key member of Saddam's government. There was evidence that he had instigated the merciless suppression of the Kurds in the north of Iraq in the 1980s, and the suppression of Shi'ite insurrection in the south during the early 1990s, the insurrection that had led indirectly to Daoud's removal from the police force. Chemical Ali's real name was Ali Hassan Abd al-Majid al-Tikriti but the nickname had taken root after he had ordered the use of mustard gas, sarin and other toxic chemicals against Kurdish-populated areas. Hundreds of thousands were thought to have died in the campaign, acts of genocide and crimes against humanity which eventually led to his conviction by the Special Iraq Tribunal and his execution in January 2010.

When Daoud enquired about his son at the gates to Chemical Ali's old home, he was rebuffed by the guards. They refused to have anything to do with him. They wouldn't let him inside and wouldn't let him speak to anyone. It didn't help that Daoud could not name the officer fully. 'Lieutenant Mike' was insufficiently precise. The guards either didn't know who he was or didn't care. Daoud waited outside the gates regardless hoping that someone would pass whom he would recognise. But no one appeared. Eventually he gave up and went home intent on returning the next day.

Nothing changed when he resumed his vigil on the Monday morning. After several more hours waiting, it occurred to him that his son might have been taken to a different location. He knew that another base in central Basra was also used by the British Army. This was Battle Group Main, although Daoud didn't know its name. It was the place where Saddam's secret police used to be housed, the place called, with black wit, the 'hospitality' centre. When he spoke to the guards at the entrance to the compound, he was ordered to leave. They had no information to give him and no intention of making any enquiries for him.

Once more he returned home. He tried calling the number Lieutenant Crosbie had given him on a slip of paper but he couldn't get a response. It was as though his son had become a ghost, invisible, lying somewhere in the buildings that used to house those agencies of Saddam that were so expert in disappearance.

18

THE SUSPENSION OF FAMILY LIFE lasted four days. Even though there was no specific reason to fear for Baha's safety, there was still mounting unease. His two young sons couldn't understand why their father had gone. No one could. It didn't make sense to them. The whole family could do little but wait.

On Thursday 18 September, Daoud Mousa and his family were still gathered, anxious, in their home in Basra. The one concern was Baha's whereabouts. What had happened to him? What could they do?

Then at 2.30 pm, according to the SIB's log, a small convoy of RMP armoured vehicles pulled up outside their home. It was a military-style operation. The whole area was quickly secured by British soldiers, fearful of attack. This was the way of things in Basra then. Nowhere was considered safe. Any dismount from vehicles had to be accompanied by defensive measures that invariably looked hostile. Soldiers with rifles watching everyone with suspicion. It was intimidating.

SSgt Daren Jay had the duty to inform the next of kin about what had happened to Baha Mousa. He was let into the home, into the main living area, where the extended family sat crowding the chairs and sofas. Jay was direct. Baha was dead. He had been in custody. Now he was dead. He had been killed. Daoud said later that the word 'torture' was used. There was an investigation under way, they were promised. Those responsible would be caught and held to account. The first step was to confirm the identity of the deceased. Daoud would be required to go to the mortuary, to see his son's body.

Baha's sons and nephews, the children who were dependent on him, the whole family gathered in that large living room listened as Daoud tried to translate what he had been told. He said later that the children understood.

No matter how ruptured, a corpse needs more than a suspected name, more than a tag fitted to a big toe, a bracelet or a certificate, to be truly claimed. It demands personal recognition, one final visual confirmation of an existence now expunged. Acquaintance is insufficient. Familiarity with the contours and clefts of a physical form are essential to penetrate the disguises of the inanimate body. However much one seeks the person in the clues of a face, identity can be elusive. Eyelids closed can transform features so that confident identification becomes impaired. And in the absence of conviction, a body cannot be buried, an investigation can be stalled, grief suspended.

The prerequisite of intimacy, the need for certainty, meant SSgt Jay had to ask Daoud Mousa to accompany him to the hospital mortuary to see the body of his son. There was no one else.

The journey was blurred for Mousa. Later he couldn't remember whether he went on his own or another member of his family went with him. But he can recall being taken into a room where a body lay beneath a white shroud. The coverall was lifted and the body beneath was exposed. Daoud Mousa said it was naked; nose bloody and broken; bruising across face and body; deep gashes around the wrists and ankles; that was what he saw. His son.

He wanted to take the body away – his son – but he was told he had to wait until the post-mortem had been conducted. A doctor was coming from the UK. It would take a few more days, probably four. He demanded an Iraqi doctor to be employed as well, to confirm the cause of death. He still had that much self-possession. His trust for the British had evaporated. And he wanted to know what had killed him – his son – what these devastating wounds about the body meant. 'Torture' was such an ugly word. It turned the fact of death into a story of pointless suffering. He had to know what was signified by those injuries, what had been endured and why.

With formal identification complete, Daoud Mousa was taken home and left to relay to his family what he had seen. The ritual of lamentation had to be honoured; that was a duty. It could not be concluded without the body. The grief had to be sustained. And surely it was amplified by what Daoud Mousa knew. He may have been a man used to violence in his professional life; as a long-serving investigative police officer he must have been equipped with the ability to distance himself from the cruel facts of victimhood. But this wasn't an anonymous casualty whose fate had to be reconstructed through a patchwork of disparate evidence. It was his son and he knew things that must have accentuated the anguish. He had seen his son, an innocent, arrested; he had heard the assurance that everything would be fine; he had seen his son placed in the care of an authority supposedly steeped in discipline and respect for the laws of war; he had seen him taken away secure in armed vehicles to be kept in an army establishment where random threat should have had no place; he had *seen* him taken away to safety; and he had seen the body in the aftermath of death. With that evidential certainty he must have had the capacity to imagine his son's last experiences. He would have been able to envisage what those moments must have been like. The state of his body gave him limitless signals to provoke that imagining, to visualise the suffering that ultimately ended with a hopeless struggle to breathe, to live. It was a legacy to foster nightmare *and* fury.

1 9

I F THE VICTIMS WERE NOW defined, the crime certainly was not. Was it murder or the application of excessive force in preventing an escape, which was the story related by Captain Moutarde in his official report to the commanding officer of 1QLR? Or was it simple assault? And was it a joint enterprise, a coordinated series of attacks on helpless individuals, or the actions of one man? Nothing was clear.

The post-mortem was awaited, of course. In the hiatus the investigators had to remain open-minded. Even without the death they had the living victims who sported some evidence of abuse. How bad that was remained a little blurred. When SSgt Jay spoke to the medics at Camp Bucca, he was told that the seven detainees still held there had only suffered minor injuries. But the two men taken away with alacrity on their arrival presented a slightly different story. Their injuries were more substantial, life-threatening in one case. Along with the death of Baha Mousa these were enough to suggest multiple offences had been committed within a British secure base over a prolonged period of time. That was unlikely to have been undertaken by one person. Nor, surely, could it have gone unnoticed.

The significance of these factors was beginning to dawn on the investigators, at first with an eye on the logistical implications: how would they be able to undertake all the necessary work – all those interviews? The resources required would stretch their limited capacities.

They would also be concerned about a second aspect, one which the staff officers of Battalion, Brigade and Division were quickly appreciating. It was already apparent that a number of soldiers must have been involved directly in physical assaults. This raised the question as to whether the whole structure of command could be implicated. For how could such treatment go unobserved? Had the abuse been sanctioned or, God forbid, ordered? Or had it somehow been permitted to occur through acquiescence or lack of care? Either way the fear of a bad press back home, which worried Major Fenton, Brigade chief of staff, would be insignificant compared with an accusation of officer complicity. The publicity would be horrendous.

The first group to prepare themselves were the military lawyers back home. The information about the death of Baha Mousa had been communicated to the Permanent Joint Headquarters (PJHQ) in Northwood military base, a few miles south of Watford in the UK, on 16 September soon after the news had travelled to Divisional headquarters in Basra. Maj. Gen. Lamb had had a videoconference with the chief of Joint Operations, Lieutenant General Sir John Reith, providing him with as much detail as he could. The information was passed on to Whitehall. A host of government personnel and politicians were sent a note on that day. The Secretary of Defence Geoff Hoon, Adam Ingram (the Armed Forces minister) and various private secretaries were told of the 'death of one detainee while in UK custody'. A serious injury to an Iraqi child was also reported in the same memo; a thirteen-year-old boy had been shot in the stomach by a guard who appeared to have 'negligently discharged' his weapon, the bullet entering the abdomen, the note said, passing through the small bowel, and possibly severing the child's spinal column and leaving him permanently paraplegic. 'The issue of compensation will be looked at separately', the memo went on to say, and 'associated defensive press lines' were attached, which the Secretary of State was invited 'to note'. Eventually a soldier in the King's Own Scottish Borderers would be prosecuted for wounding the boy. Private Alexander Johnston was convicted of the offence in 2005. His sentence was a fine of £750 and compensation of £2,000 to be paid to the family. But the shooting of the young boy had been at a different base and was a wholly unrelated incident. The death of Baha Mousa was presented as the first and perhaps more damaging story.

The memo to various levels of government servants and politicians played on the tale that Baha Mousa had been detained as part of 'an anti-terrorist and anti-criminal arrest operation'. It talked of the confiscation of small caches of rocket-propelled grenades and other weapons. And then it related how Mousa had 'repeatedly tried to escape and also allegedly lashed out at guards'. It was in this context, the memo said, that 'two members of the guard restrained him and replaced his hood. His pulse was also apparently checked at this time. Three minutes later the guard suspected that he might not be breathing.' And then the failed resuscitation attempts had occurred. The note was specific too about the length of time Baha Mousa had been hooded: twenty-three hours and forty minutes. No mention was made, though, about the questioning process or preparatory disorientation, but more information was promised as soon as it became available.

Now the lawyers were examining the issues raised with some care. The legal corps had been experiencing considerable pressure since the beginning of the Iraq War. They were part of the military intention to operate a system of 'heavy lawyering'. This was the phrase used by Air Chief Marshal Brian Burridge when asked in a parliamentary committee session to say how the forces prepared themselves to respect international laws of war, the Geneva Conventions in particular. It meant that packs of lawyers would operate at every level, from the front line to the UK, he said. They would all be called upon to advise on what weapons to use, what targets were permissible and how to treat prisoners and civilians. At PJHQ, the legal team were asked constantly to approve certain strategic actions and to comment on individual cases; the lawyering was indeed heavy for them.

Lieutenant Colonel Nick Clapham was Britain's senior legal link with Iraq. He was an experienced legal officer. He was called into various meetings on 16 September, soon after the death of Baha Mousa had been reported up the chain of command. The subject was the death in custody. More pointedly it focused on the practice of hooding detainees. The original story unearthed by Captain Moutarde back at Battle Group Main that the dead man had his head covered by a sandbag at the time of his death had caught people's attention. But there wasn't an automatic condemnation of the practice. There was some vacillation. At least, Clapham sought the advice of a medical officer at HQ to clarify why hooding might not be a good thing. He was told that the use of hoods could cause breathing difficulties that might induce asphyxiation. The medical opinion would hardly have been necessary if it was beyond doubt that hooding was outlawed.

Lt Col Clapham nevertheless thought it wise to issue written email advice to a host of HQ personnel. He repeated the medical opinion he had received. And then he confirmed that for a limited period *blindfolds* could be used, or in his language were 'legally sustainable', to stop a detainee seeing other prisoners or militarily sensitive areas. If they were used as part of 'disorientation', to make the prisoner more disposed to provide information during interrogation, this was *not* tolerable, although strangely he recognised that 'this may prove controversial and contrary to standard practice'. He wrote that the International Committee of the Red Cross had already told the army that hooding was unlawful. They had visited Camp Bucca earlier in the year and had found prisoners blindfolded. That was a breach of the Geneva Conventions, they had said. Clapham thought any disorientation by depriving a prisoner of sight would more obviously be a contravention. He recommended hooding should cease and alternative means used – blackened ski goggles, for instance.

Lieutenant Colonel Ewan Duncan was deputy head of intelligence in Baghdad at the time Clapham issued his advice. He read the note and replied quickly, believing that the ending of hooding, as suggested by Clapham, would have 'far-reaching implications'. He pointed to a difference of opinion with the US and possible 'adverse impacts upon interrogations', one of which would be, he wrote, 'UK involvement in US ops where blindfolding is the milder end of the spectrum'. He didn't specify what the not so mild treatment favoured by the Americans might be. And nor did he condemn it. But there was an implication that the relationship with Britain's military partner might be badly affected if the legal advice was followed too strictly or even at all. He also thought that a ban would impinge on 'current doctrine and teaching' at the British Army's intelligence training facility back in the UK. Nothing explicit was mentioned about this either but the inference was clear: the legal position was neither widely recognised nor appreciated.

Whilst this correspondence was developing, the Division legal commander at Basra airport, Lieutenant Colonel Charlie Barnett, was being drawn into the plethora of communications. He had senior responsibility for ensuring British forces in south-eastern Iraq were acting lawfully. He sent round an email to the British brigades in south-east Iraq on 17 September. His recommendation did not cavil about hooding: 'My strong advice has been that all hooding be immediately stopped', he wrote in bold letters. He was adamant that they were 'responsible in law for the well-being of any individual in our custody'. Using sandbags for hoods, indeed depriving someone of their sight except for necessary security reasons, was not conducive to that obligation. He believed that this was already understood. Barnett worried that 'corporate knowledge had been lost'. If he was right, who had lost it? And what exactly had been lost?

20

ALTHOUGH MANY MILITARY AND GOVERNMENTAL *personnel would later swear ignorance, the British Establishment and the British Army had much more than 'corporate knowledge' about hooding and disorientation as a part of a regime of interrogation. They had intimate experience of these matters and had been mauled politically and legally for that familiarity. None of this arose from a war in some far distant ex-colonial back-water. It was much closer to home than that.*

In August 1971, the Northern Ireland administration of the time, with the support and direction of the British government in London as well as the British Army, introduced internment in the province as a means of trying to combat the perceived or actual threats posed by the Irish Republican Army, the IRA. Mass detention without trial was believed to offer some kind of solution to the incipient but increasingly horrific violence that the British were impotent to counter. The policy would allow the authorities to trawl for suspects, to interrogate them and obtain information about a group and a movement which so easily merged into the housing estates of Belfast and Derry. Intelligence was simply not being acquired by normal police methods. Internment, swift and extensive, was conceived as a way of shocking the republican community to reveal its secrets. And if that didn't work, at least those people suspected of supporting the growing number of bombings and shootings and backstreet punishments would be out of action for a while.

As soon as the new internment regime was authorised, the Northern Ireland security forces launched Operation Demetrius. There was no explanation why that name was chosen. It didn't signify anything. Perhaps there is a database of official titles that are picked in some form of alphabetical order, like hurricanes. But at 4 am on Monday 9 August 1971 the army was issued with a list of over 400 names to pursue and arrest. There was close coordination under police guidance and a slew of squads swept up nearly 350 people on that list. It was a massive operation. Those arrested were initially taken to specially prepared holding centres in central Belfast, Londonderry and County Down. Some were released fairly quickly. The majority were later transferred to other detention facilities for further interrogation.

Operation Demetrius was not the end of the initiative. Internment orders continued to be issued by the Northern Ireland authorities until the British government assumed direct rule over the province in March 1972. More than 3,000 people had been processed through the system by that time.

It didn't take that long, though, for internment to become known as a ruth-less and brutal exercise. Quite apart from the indiscriminate nature of the arrests and detentions and the complete dismissal of any rights normally attaching to someone taken into custody, internees would reappear many days later, released without charge, crying out about the ill-treatment they had received in the detention centres. Persistent allegations filtered through to the news agencies and on towards the government of Ireland.

The stories were remarkably similar. Many spoke of a systematic handling insidious in its ability to induce hopelessness, fear and chronic stress. Five methods of treatment were prevalent across the detention centres: hooding with black or navy blue pillowcases, kept in place for prolonged periods of time except when the person was being interrogated; being forced to stand, spreadeagled against a wall; constant loud and hissing noise; being prevented from sleeping before inter-rogation; and being fed and watered only in the most meagre way. The measures were intended to break the internees, to make them more susceptible to inter-rogation so that in their weakened and disoriented state they would be more amenable to answering questions.

When all this information came to public notice, the British government announced a committee of inquiry to investigate the reports. Its findings were made public in November 1971. There was an outcry; contrary to the evidence, the report concluded that the five techniques did not amount to brutality.

The Conservative administration under Prime Minister Edward Heath quickly felt a scandal brewing. Another report was commissioned. But this helped little when it was released in January 1972. The Parker Report, named after its chairman, Lord Parker of Waddington, presented a majority conclusion which accepted that the techniques were illegal according to English law but suggested that they need not be wholly ruled out on moral grounds. They might still be acceptable if certain recommended safeguards were adopted.

Heath and his ministers found even this unpalatable. Could they really maintain support for such a clearly abusive approach? The very day the report was released Heath made a statement to the House of Commons designed to forestall any political uproar. He said that despite the reports commissioned, the government had decided that the five techniques that had come to light 'will not be used in future as an aid to interrogation'. But he didn't stop there. He then sought to bind any government that followed his.

'The statement that I have made', he said, 'covers all future circumstances.' If a subsequent government should ever change its mind on this policy, he went on, then they 'would probably have to come to the House and ask for powers to do it'.

This wasn't all. The government issued plain orders to both the army and police that any arrest and interrogation had to be in accordance with humane treatment and that all five techniques were forbidden. A directive issued by the Joint Intelligence Committee in 1972 said:

Searching and sustained interrogation should be carried out in a disciplined atmosphere, and it may in some circumstances be necessary for interrogation to be carried out by night. But no form of coercion is to be inflicted on persons being interrogated. Persons who refuse to answer questions are not to be threatened, insulted, or exposed to other forms of ill-treatment. Techniques such as the following are prohibited –

(a)   Any form of blindfold or hood;
(b)   The forcing of a subject to stand or adopt any position of stress for long periods to induce physical exhaustion;
(c)   The use of noise-producing equipment;
(d)   Deliberate deprivation of sleep;
(e)   The use of a restricted diet to weaken a subject's resistance.

Proper arrangements should be made for the physical needs, including food and drink, of all persons being interrogated.

The directive was also explicit about command responsibility. It said control of interrogation in accordance with these rules 'must rest with the senior British Service Commander in the country concerned'.

It couldn't have been clearer. Even so, when the case against the British government for the abuses committed during internment was finally heard in front of the European Court of Human Rights in 1977, the UK's Attorney General confirmed once again that 'the "five techniques" will not in any circumstances be reintroduced as an aid to interrogation'.

Hardly surprising, then, that decades later there might be some sensitivity about hooding and disorientation as a prelude to interrogation. Strangely, though, the concerns worming their way into the emails and briefings from army HQ to Whitehall offices immediately after Baha Mousa's death forgot the permanent order issued in 1972. In fact, no one mentioned the Northern Ireland case. It might have been a standard judgment explored in most human-rights textbooks and basic law-school courses, but that knowledge seemed to have slipped from institutional consciousness.

21

Modern warfare for the British Army is like modern politics: substance may matter but so does image. A poor press review, an unsympathetic storyline, a condemnatory article can hurt. For the higher military echelons it tastes like defeat; and a defeat that can lead to long-term damage: budget cuts, reduced influence, embarrassment.

The army couldn't afford bad publicity when it came to Iraq; they were already labouring under a globally unpopular intervention. To add to that weight of scorn was the last thing they needed. A media operations division was established in Basra immediately after the invasion to meet the demands of the insatiable industry of news and to keep some control over the truth of the occupation. Tens of liaison officers, communication experts and support staff, drawn from across the services, were brought together to facilitate the army's media management. They were based at Basra International Airport.

On 17 September, the officer commanding the media operations unit informed an RMP officer at Division, Major West, that three of his drivers had seen something at Battle Group Main a couple of days before. Major West quickly telephoned Captain Gale Nugent at the SIB with their names. They promised to be the first uncompromised witnesses of events in the detention facility where Baha Mousa died. It was a breakthrough in the making.

Captain Nugent dispatched one of her investigators to interview the three men the next day. Sergeant Birch drove to the airport and managed to track down only one of them: Lance Corporal James Riley.

Riley had joined media ops in June 2003. He'd been posted as the transport manager and assigned responsibility for driving TV and press crews about the region. It soon became apparent that the evidence he had to give would be crucial. Although Riley couldn't speak to Sergeant Birch for long, he was able to make a brief statement about what he had seen.

A *GMTV* news team had been visiting the region between 14 and 16 September. Cordelia Kretzschmar, a young correspondent who had only recently joined the programme, was making her name, taking the opportunities offered by the army to satisfy that increasing desire for 'embedded' reportage. On this occasion, though, she was only to be shown around, to meet some of the commanders in the field, trying to construct a story about life in Basra now that the war was over. She wanted to interview some Iraqis to discover how life had improved now that the British were in control. She had a cameraman and sound recordist with her.

The obvious place to take the *GMTV* crew was Basra Central. That meant visiting Battle Group Main given that 1QLR were responsible for security in the city. The plan was to take the reporter to see the commanding officer to obtain some background information.

Riley was one of the drivers assigned to carry the *GMTV* team into Basra. They arrived at 1QLR's compound in two 'snatch' Land Rovers at about lunchtime on Monday 15th. It was quite a gathering; apart from the TV people there were two drivers, Riley and Lance Bombardier Richard Betteridge, a 'top cover' (a soldier who had a weapon prepared and drawn, ready to fire, watching from the roof of a vehicle to respond to any threat), Senior Aircraftman Scott Hughes, and a couple of chaperoning officers.

The Land Rovers pulled into Battle Group Main, drove the few hundred yards down the dusty compound road from the main entrance and parked outside the main accommodation block. They had pulled up about twenty yards from the detention facility, where a small group of soldiers were hanging about, sitting, ambling, lying in the midday heat and glare.

The officers immediately escorted the TV reporter and crew to see the commanding officer. Riley, Betteridge and Hughes were ordered to stay put in case they had to leave quickly. It was a normal precaution.

It was also normal for troops standing down within an army facility to unload their weapons. They were supposed to do so somewhere safe. The designated site at Battle Group Main happened to be next to the detention facility at the walled edge of the compound. Soon after arriving all three men went to make sure their rifles had no ammunition in their chambers. Riley told Sergeant Birch that as he went to unload his weapon he walked past the detention block and looked into the doorway. A number of men, obviously Iraqi, were sitting cross-legged on the floor, in a line, heads covered by sandbags, hands tied, palms together as though in supplication, arms stretched out in front of them, shoulder height. There was a lot of shouting; loud, like an irritated teacher. 'No sleep, Grandad,' Riley heard.

Riley returned to his vehicle, putting his rifle back into the Land Rover.

He recalled to Birch that he had seen three British soldiers at the detention block that he would recognise again: two youngsters, about twenty and eighteen, he thought, wearing 1QLR insignia, and a third, older than the others, in his thirties, stocky, brown-haired, gap-toothed, about five feet ten, dressed in standard desert issue fatigues and boots. He appeared to be the one in charge. He didn't have 1QLR identification badges, but the way he carried himself, the physical presence he emitted, suggested that he was the man giving the orders.

Leaning against the Land Rovers waiting for the TV people to return, Riley saw the oldest soldier bring one of the sandbagged men to the Portaloo, which was set against the outside wall of the detention block. The soldier shouted continuously, telling the man piercingly to 'hurry up, hurry up', and after a few minutes, maybe less, to 'get out, get out'. When the man emerged from the toilet, the soldier wound a plastic line around his wrists and led him back into the block.

Perhaps after ten minutes or so, curiosity overcame all three of the media soldiers. They walked over to the facility to see what was going on. Riley told Sergeant Birch that he looked inside one of the doorways and saw a sandbagged detainee, sitting cross-legged on the floor, facing the wall. He had his arms outstretched with his wrists bound by cable. Riley saw him fall backwards and lie still. The stocky, gap-toothed soldier came back into the room and began shouting again, 'get up, get up', and forcing him back into position. The man fell again and the soldier shouted again. It was almost ritualistic.

Riley heard one of the other guards say 'Look at this', indicating the other room in the block. Riley looked inside. There were four more detainees there, heads also covered with sandbags, facing the wall, cross-legged, arms out in front of them. The stocky soldier entered the room and announced 'This is the chorus', laughing. With that, the stocky soldier lined the prisoners up and gave a kick to each one. And with each kick came a groan or a shriek at a slightly different pitch: this was the 'choir'. The soldiers giggled at the macabre joke.

Riley and Betteridge walked away, back to the Land Rovers. SAC Hughes, Riley said, stayed at the detention block, talking with one of the younger guards. He was there for about twenty minutes until he also came back to the vehicles. The *GMTV* crew and officers returned, the drivers prepared their weapons, and they all left in convoy. Neither Riley nor the other two said a thing as they drove away.

LCpl Riley didn't have time to say any more to Sergeant Birch. He was called away to return to his duties. But Birch had enough. He brought the account back to the SIB. As Captain Nugent and her colleagues had hoped, it provided the first shard of testimony about the abuse *as it was happening*. It substantiated some of the complaints made by the detainees, as well as explained their injuries. It was a significant development, as police officers are fond of saying. Here was an independent witness, untainted by association with the suspect soldiers of 1QLR, who had seen maltreatment *and* could probably identify the men involved. And there were two other witnesses who could presumably verify Riley's testimony. It promised to be a perfect triangulation of evidence which prosecutors adore.

It might have been the breakthrough the SIB needed, but Riley's story betrayed a double irony – and a tragedy too. The irony? Here was a media-operations engagement taking a TV film crew almost unerringly straight to within spitting distance of a story that would have instantly wrecked the army's whole media strategy. And here too was a TV film crew searching for a story parking a few yards away from one of the most shocking revelations of the Iraq War. And missing it.

The tragedy? That chance visit could have saved Baha Mousa's life. If the *GMTV* crew had had an inkling of something happening in the facility, if they had heard a cry or a shout and had been intrigued, if the media drivers had said something to their officers when they had returned from speaking to the CO, even spoken in front of the glamorous reporter and her technicians, then it's hard to believe Baha Mousa would have been killed as he was. Their visit had been several hours *before* he received his final fatal beating. An intervention on that Monday lunchtime would surely have stopped the abuse there and then.

But that didn't happen. Nothing was said until the next day. And then it was too late.

22

LCPL RILEY'S INFORMATION, INCOMPLETE AS it was, confirmed some key details for the SIB team. They were certain now that assaults against the detainees, presumably including Baha Mousa, had been going on for some time before his death. The media drivers had seen the prisoners in such a poor state at lunchtime on the Monday (at least twenty-four hours after they had been arrested on the Sunday morning) that it seemed likely that they had been treated in this way for a while. And clearly a number of people must have been involved either as direct perpetrators or as witnesses. These were the ones they now had to find.

SSgt Sherrie Cooper had already confirmed at her first briefing with Captain Moutarde, the adjutant, and Lt Col Mendonça that Baha Mousa had been picked up by one of the battalion's companies during a raid on a hotel on Sunday 14 September. The men of that company, A or Anzio Company as it was called, were the arresting troops. They were divided into two 'multiples' under the overall command of Major Englefield. The first, commanded by Colour Sergeant Christopher Hollender, had been responsible for securing the perimeter of the hotel during the operation. The second, commanded by Lieutenant Craig Rodgers, carried out the arrests inside.

There were sixteen men in Lieutenant Rodgers' multiple. These men had also guarded the detainees at some point. Sergeant Colley's notebook confirmed that those soldiers he had found crowding about the detention facility after Baha Mousa was pronounced dead were from that multiple. He had listed them:

Lieutenant Rodgers
Private Christopher Allibone
Private Thomas Appleby
Private Gareth Aspinall
Private Peter Bentham
Private Aaron Cooper
Corporal John Douglas
Private Lee Graham
Private Jonathan Hunt
Private Damien Kenny
Private Stuart MacKenzie
Private Garry Reader
LCpl Adrian Redfearn
Private Paul Stirland

All of these were potential witnesses, perhaps even suspects. They had to be seen and questioned. (There were two others in Anzio Company who hadn't been identified by Colley that night: Private David Fearon and Fusilier Richards.)

Whilst SSgt Jay was informing Daoud Mousa about his son's death and obtaining confirmation of the deceased's identity, and Sergeant Birch was seeing James Riley and taking his first statement, and Captain Nugent was following up the appointment of a British-based pathologist to undertake the post-mortem, SSgt Cooper headed back to Battle Group Main to start questioning the Anzio Company soldiers. The detective who was to take the investigation over from her in a few days, WO David Spence, went with her.

The SIB officers had prepared for their task by constructing a pro forma questionnaire to hand to the soldiers. It was a way of ensuring that they didn't forget important questions and the interviews could be dealt with quickly. But it would also implant an insidious seed in the soldiers' minds that the SIB already knew what had happened. Honesty and candour might result. It was worth a try.

Apart from basic background information, the form asked how the guards had been briefed on dealing with the detainees; it asked what guard duty they had performed at the detention cell, who had been there, whether they had given the prisoners water, food, toilet breaks. It asked whether they had applied stress positions, hoods, plasticuffs; it asked whether any force was used against the prisoners, whether they had 'corrected them', used hand, boot, rifle; whether any complaints had been made about their treatment. It asked whether visitors had entered the detention block, whether any of the detainees were taken away during their guard duty and if any injuries had been seen when they had been returned. And finally, 'Did you strike any of the detainees whilst they were in your care?'

Anyone filling out the form would have known precisely where the investigation was going. It wasn't intended to be subtle.

Armed with the form, Cooper and Spence, along with a couple of other RMP officers (sergeants Gordon and Robinson) began to see the men of Anzio Company. They worked in pairs. Sergeants Gordon and Robinson started with Private Aaron Cooper, SSgt Sherrie Cooper and WO Spence with Lieutenant Rodgers.

Almost before they had a chance to begin, Sergeant Gordon rushed from his interview with Private Cooper and interrupted WO Spence's session with Lieutenant Rodgers. Gordon said he had some information.

He needed help. Private Cooper had admitted that he'd been in the detention centre with the detainee at the moment he had died. He'd said LCpl Redfearn had been there too. Sergeant Gordon wanted to know what he should do.

The SIB team quickly conferred. They decided they couldn't continue the interviews. Private Cooper hadn't been cautioned yet. He had no legal representative. Anything he said that was incriminating might be unusable. It could be thrown out of court, tainted for failing to follow proper procedure. They had to stop. Private Cooper had to be arrested, Redfearn too, both on suspicion of involvement in the death of Baha Mousa.

Sergeant Gordon was sent to arrest Private Cooper. The interview with Lieutenant Rodgers was halted. They called for LCpl Redfearn and promptly arrested him as well.* The whole interviewing process ceased. They bundled the two suspects into their vehicles and headed off to SIB HQ. The two suspects were to be cautioned and a legal advisor appointed. The investigation had suddenly acquired momentum.

23

THE SPECIAL INVESTIGATION BRANCH OF the Royal Military Police based in Basra was now deep into its third day of inquiries. The team needed to take stock. Staff sergeants Cooper and Jay, WO Spence and Captain Nugent sat down to consider the current state of their investigations.

They must have been fairly pleased with themselves. They had an independent witness in LCpl Riley with the prospect of two others to follow (SAC Hughes and LBdr Betteridge). They had initial statements from the surviving detainees and photographic evidence of their injuries. They had a post-mortem arranged and photographs of the state of the corpse, which already indicated a massive assault had occurred sometime prior to death. And now they had two men of Anzio Company, Private Aaron Cooper and LCpl Redfearn, placed by their own admission at the scene when Baha Mousa had collapsed and died. Whether these soldiers were complicit in

---

* It seems strange that SSgt Cooper hadn't taken the same step when she saw Corporal Donald Payne a few days earlier and he had admitted that he was restraining Baha Mousa at the time of the detainee's death.

the death, assailants or material witnesses had yet to be established. But the shape of the case was beginning to form.

The next step was to interview Private Cooper and LCpl Redfearn under caution, to treat them as suspects. An RAF legal advisor, Flight Lieutenant Hughes, was driven over from Division headquarters in Basra airport to speak with the two men. He took so long taking instructions from Private Cooper that they decided to postpone any further action until the following day, 19 September. Cooper and Redfearn were taken to separate quarters and left alone for the night.

Late the next morning, Sergeant Gordon of the SIB was assigned the task of interviewing Private Cooper. A room with a portable air-conditioning unit and a tape recorder on the desk was provided. All interviews had to be undertaken in pairs so SSgt Sherrie Cooper sat in to help with particular questions if needed. They had arranged for Sergeant Palmer from 1QLR to be there. He was to act as the battalion's observer, but there was nothing really for him to do other than watch. And, of course, Private Cooper was present, recently cautioned, with Flt Lt Hughes accompanying him as his legal counsel. The room was crowded.

Questioning began.

For an hour Sergeant Gordon doggedly asked questions about the detainees. When were they arrested? What happened at the hotel? What happened when they returned to Battle Group Main? Who did what to whom?

Private Cooper's responses were repetitive.

Gordon asked 'Yesterday afternoon I arrested you on suspicion of being concerned in the death of Mr Baha Mousa, is that correct?'

Cooper said 'No comment.'

Gordon asked 'OK, at 17.35 hours you was then taken by myself to the cookhouse here on camp where you was fed, had a drink and returned to, once again, to 61 Section SIB at 18.10 hours, is that correct?'

Cooper said 'No comment.'

It was exasperating. Sergeant Gordon explained to Flt Lt Hughes and Private Cooper that all he wanted to do was confirm for the record that Cooper had been treated well since his arrest. It was hardly controversial.

Cooper said 'No comment.'

And so it continued. More than 170 times Private Cooper responded to the questions put to him with 'No comment.'

Sergeant Gordon's perseverance was remarkable. He went through every aspect of the case in sequence, from the arrest of the detainees up to the moment of Baha Mousa's death. Every so often he would turn to SSgt Sherrie Cooper and ask her 'Anything up to now, Sherrie?' There wasn't. The only interruption to the monotonous flow was when they had to change tapes. And when Sergeant Gordon had an attack of cramp. They all laughed about that. Gordon had tried to avoid crying out in case it would be misinterpreted when heard on the tape. Police brutality. But eventually he had to explain. Flt Lt Hughes joked 'Don't die on us.'

Sergeant Gordon finally reached the end of his list of prepared questions. He had made not the slightest impact. By now it was becoming increasingly and uncomfortably cold in the room, freezing in fact. The air-conditioning unit was rudimentary, loud and a little too effective. It produced an atmosphere of icy unreality with the desert visible through the closed windows. Gordon felt an infuriating pressure. It all seemed so futile. Forty-five minutes of question after question with the same response time after time.

But Sergeant Gordon had one final query. Would Cooper be prepared to take part in an identity parade? He offered a long break for Cooper to think about it, to have a chat with his advisor.

The question may have seemed innocuous, but it was pregnant with implication. If there was to be an identity parade then it meant they had an eyewitness. Maybe more than one. Maybe someone who was in the detention block when Baha Mousa had died, someone who wasn't supposed to say anything, perhaps.

Private Cooper and his legal advisor were given a private room. What could they have talked about? There was a man dead. Someone had to be responsible. Cooper had been identified as there at the death. How would that look? This was potential homicide, not some regulatory misdemeanour. It wasn't even a matter of regimental discipline. It would involve a court martial, and then prison perhaps, and a long sentence if he was blamed and found guilty. He wouldn't just be sacrificing his army career; this would be his whole life. A criminal record to take back to civvy street. And what would people see? Murder? War crime? Where would that leave him? If he had nothing to be guilty about, then staying silent wouldn't necessarily help him. It might even drag him down. And if he *was* somehow responsible, even in part, then a lack of cooperation would hardly do him any credit.

Shortly after 3 pm everyone reconvened in the interview room. Cooper and his advisor had been talking for an hour.

Sergeant Gordon switched the tapes back on and asked Private Cooper whether anyone had applied any pressure on him during the break. It was a standard opening when there was a hint of confession in the air.

'No,' was Cooper's reply.

And did he have anything to say now?

'No . . . sorry, yes.'

OK.

'I'm not going to ask you any questions,' said Sergeant Gordon. He had spent long enough doing that already. 'If you've got something to say, Private Cooper, the tape's running.'

And Cooper launched into his story. 'On the evening, I'm not sure of the day . . .'

Before he could get any further Gordon stopped him, apologised, turned off the air conditioning unit because it had become a 'pain in the arse' (it was still making too much noise and the little room was now ridiculously cold and Gordon could sense this was going to take some time) and then settled back to listen.

Cooper spoke in rapid, anxious, occasionally interrupted, snatched monologues.

'On the evening, the Monday evening, I'm not sure of the date, I arrived to take over the rest of my call sign that was on duty guarding the detainees at 1QLR main. As I arrived at the, where the detainees were being held I heard . . . I heard someone asking for assistance. I looked in the room where six of the detainees were being held. I heard . . . there was no one there, no one asking for assistance, so I went on to another room just in between two of the main rooms and I seen an NCO with one of the detainees on the floor. He told me that the detainee had unhooded himself and taken off his plasticuffs . . . I was followed by another NCO with a torch to help with getting the plasticuffs back on as there was no electricity or lighting in the building, I uh . . .'

There was a pause, a quick breath. Cooper swallowed hard.

'Take your time,' Flt Lt Hughes said.

Cooper continued as breathlessly as before.

'The NCO who asked me to assist him had himself positioned on top of the prisoner with one knee just below the shoulder blade . . . and then

I got on the floor on my knees with one leg in between the legs of the prisoner . . . one of the detainee's legs in between my legs on the floor and I went to put on the plasticuffs . . .'

'Take your time, OK?' said Flt Lt Hughes again.

Private Cooper stopped. He had a drink of water. But there was little change in his delivery.

'We, the detainee, the prisoner, was struggling trying to get from . . . get away . . . get us not to touch him . . . he was struggling around. I tried to get the plasticuffs . . . I got the plasticuffs on to him but the detainee managed to get free once again. I'm not sure if he . . . if he forced them open himself or there was a problem with the plasticuffs. I seen the NCO stand up and as I also stood up . . . moved away from the prisoner and that . . . I seen the NCO kick, kick the prisoner twice in the right-hand side . . . from then I got the . . . we both got the, like, to the prisoner tried to restrain him again. He was still moving around trying to get away from us frantically and he hit his head, left-hand side of his head against a wall while trying to get away . . . while trying to struggle and after that I managed to . . . not long after that I managed to get the plasticuffs actually on to the prisoner and then we sat him upright against a wall. He flopped . . . he sort of fell over to the left-hand side so we put him on . . . we placed him on to the floor. I was asked then to check his pulse . . . I did check his pulse and he did have a pulse . . . another person . . . another soldier came into the room and checked him for breathing. I left the room and then I heard that the prisoner was not breathing . . .'

A short pause at last.

Sergeant Gordon asked 'Is that all?'

It wasn't.

'I went to . . . after hearing that the prisoner had stopped breathing, I went into like a little corridor just outside of the middle room. I watched as the two lads tried to revive . . . get the detainee breathing again. The NCO who was involved was stood at the side of me. He went to go and get some more, to get, get more medical attention, stretcher and I held the torch while the lads tried to revive . . . to get him breathing again then as soon as medical attention, the stretcher arrived I moved out the way and let them carry on.'

Another pause.

And finally Cooper said 'That's it.'

This time he was sure. But, for the next hour or so, Sergeant Gordon picked away at the details of Private Cooper's story. The first thing he wanted to know was why he had changed his mind about talking to them. What had happened between saying 'No comment' over 170 times and suddenly vomiting this confession?

'If I didn't speak out,' Private Cooper said, 'I could be blamed for the death of this man.'

It wasn't the result of any influence applied to him?

'No.'

Sergeant Gordon now wanted to know the identity of the other men in the small room of the detention block where Baha Mousa collapsed. Cooper told him that the NCO who had pinned Mousa to the floor was Corporal Donald Payne. The NCO who had followed Cooper into the room with the torch was LCpl Redfearn, who had only watched as Payne and Cooper restrained the prisoner. Private Reader was the soldier who had appeared a little later, who had checked the prisoner's pulse and who had tried to resuscitate him.

How had the detainees been treated? Sergeant Gordon asked. Who had briefed the guards? What were they supposed to do? Private Cooper said he was told not to let the prisoners sleep. They were to be cuffed and hooded. Although he wasn't at the detention facility all the time, when he *was* on duty, from late Monday night until early Tuesday, Sergeant Smith had come in, he said, and told him to uncuff the prisoners, take their hoods off and let them sleep. Then half an hour later, Cooper said, a colour sergeant, he thought Colour Sergeant Livesey, had entered the detention block and told them to put the cuffs and hoods back on and to get them all sitting up away from the wall. That was the way they kept them throughout the night. The guards would make a noise if the prisoners fell asleep. Cooper said they were given a metal bar to bang against the floor. But Cooper was adamant he didn't punch or kick anyone.

At 5.17 pm the interview was brought to a close. Private Cooper assented to take part in an identification parade. He was then rearrested so that his boots could be removed. The SIB team wanted to test whether they matched any marks on the detainees. That done, he was sent back to his Company.

SSgt Sherrie Cooper then told LCpl Redfearn that he was to be treated only as a witness and he too was released back to his unit.

24

I T WAS 19 SEPTEMBER, FOUR days after Baha Mousa's death. The SIB had positive identification of the soldiers who were with him when he had stopped breathing. It was a little odd that it had taken them this long, given that Captain Moutarde had delivered his memorandum to his commanding officer before midnight on 15 September naming the people who had been there. The memo was supposed to have been given to the SIB the next morning and it was certainly kept amongst the investigation files. Years later, it would be shown to Captain Moutarde, who said he'd forgotten all about it. And in that memo it stated emphatically that Corporal Payne and Private Cooper had restrained the prisoner, Private Reader had tried to resuscitate him and LCpl Redfearn had stood and watched. In short, all the names and suspected roles the SIB needed.

Was anything lost because of the delay? Obviously, the sooner a potential suspect is seen and interviewed the better the chance to avoid collusion, the fabrication of a collective story.

But in apparent ignorance of this setback, the SIB were moving forward with some confidence. Even though the task of collecting evidence remained colossal, and their limited resources and the conditions on the ground made any investigation difficult, there were definite signs of progress. It was hoped that SAC Scott Hughes, one of the media operations section who had visited Battle Group Main with *GMTV* on the afternoon of Mousa's death, would offer another crucial segment of the case. WO Spence was sent to Basra International Airport to speak to him.

The importance of what Hughes had to say became apparent immediately the interview began at 9.30 am on 19 September. He confirmed LCpl Riley's evidence and embellished it. His testimony was richer in detail but shadowed by shame.

SAC Hughes was a young man, twenty years old, a little nervous, timid even. Not only did he admit to watching the treatment of the detainees for more than an hour, fascinated by what he saw, unable or unwilling to turn his back on the scene, but he also said he failed to talk about it to either his colleagues or superior officers afterwards. He observed the violence as a voyeur, passive, neither participating nor condemning nor sufficiently appalled to walk away. The soldier he saw perpetrate most of the violence 'scared' him. And he worried that the guards had been ordered to act in that way. Any interference, he thought, might go badly for him. He liked his job and didn't want it jeopardised. If this was how prisoners were treated, he thought, then he might 'look stupid' if he made a fuss. He only became emboldened to say something back at his HQ and only *after* he had been told that a man had died in custody down at Battle Group Main. His commanding officer had mentioned the rapidly circulating rumours, thinking that they may have to go back there, although for what purpose wasn't clear, and Hughes had shuddered at the idea. He'd said he didn't want to go. When asked why, he'd said he had seen detainees being abused there. The officer had told him to report the information immediately to Major Mayo who led the media ops section, which he'd done. It had been Major Mayo who had informed Captain Nugent of the RMP.

WO Spence wasn't interested in the moral ambiguities of SAC Hughes' character. He wanted facts.

Hughes told Spence all he had seen and heard. He repeated Riley's account of how the media ops crew had found itself in Battle Group Main with the *GMTV* team. He said he was first drawn to the shouting at the detention block soon after arriving in the Land Rovers. LCpl Riley heard it and wondered what was going on, he said. Hughes was intrigued and used the excuse of unloading his weapon to pass the block on the way to the designated unloading bay. As he did so, he saw a soldier, whom he described as Male 1, next to the Portaloos outside the detention block, waiting with a set of plastic handcuffs, tapping them against his hand, saying to someone inside the temporary toilet 'Come on, you can piss for Iraq inside – why can't you do it now?'

'Yes sir, yes sir,' an Iraqi voice came from the toilet.

Hughes looked in both doorways to the detention facility and saw various prisoners, cuffed to the front, sandbags over their heads and sitting cross-legged on the floor. He passed two other British male soldiers slouched on the ground outside one of the doors and greeted one of them. He said 'What's going on here then?'

The second soldier, 'baby face', about eighteen, told him they'd just caught the prisoners in a hotel with a load of weapons and they were suspected of killing the three MPs in August.

A fourth soldier emerged from the detention block. Hughes went on with his chat.

'How can you do this?' he asked, not accusatorily but with disgust at the smell of sweat and piss coming from the building.

The second soldier shrugged. The fourth said 'If they caught *you* they'd cut your balls off and make you eat them.'

Hughes wandered back to the Land Rovers, hearing Male 1 shouting all the while 'Get your fucking arms up, keep that fucking head up! I'm gonna fucking kill you!'

He put his rifle in one of the vehicles and then returned to the detention block, like a moth to flame.

With his usual friendly demeanour, Hughes continued to chat with the guards. He asked them how long the prisoners had been in the detention block.

'Thirty-six hours,' he was told. 'They're being interrogated one by one. We've been ordered not to let them sleep.'

'Why does it stink so much in there?' Hughes asked.

'They've pissed and shat themselves. Wouldn't you, with bags on your head and all that shouting?'

Male 1 came out. He asked Hughes what he was doing there and Hughes told him about the *GMTV* crew. Then Male 1 went back into the block and resumed his shouting, 'Get your fucking head up!' He didn't seem bothered that journalists were sitting having tea in the CO's office a hundred metres away.

Hughes peered through the doorway. He saw Male 1 behind one of the detainees, placing his hand over the hooded man's face and yanking his head back.

The scene drew Hughes into the building. No one objected to his entering. The guards were wholly at ease with his presence. He said hello in Arabic to one of the inmates who responded in kind. To another he asked, again in Arabic, 'How are you?' These were stock phrases learnt by army personnel. The detainee placed his hands on his heart and then his forehead in salutation.

Hughes wandered along the corridor between the two main rooms of the block and looked into the middle section that was an old toilet. There was a man there sitting on a flattened cardboard box. Hughes carried on ambling from room to room, back and forth.

'Get your fucking arms up, Grandad!' Male 1 again.

For an hour, maybe an hour and a half, Hughes said, he meandered about the building, inside, outside, watching through the windows, the doorways. He saw many things, he said: Male 1 pressing his hand over a sandbagged face seemingly trying to stick his fingers in the man's eye sockets; the soldiers kicking at the feet of those inmates who didn't keep their legs crossed beneath them; the fourth soldier instructing one of the detainees to say 'Fuck you' whenever he clicked his fingers; the same guard squirting water hard into the mouth of a prisoner who had pleaded that he was thirsty, so hard that he couldn't swallow it properly; Male 1 shouting 'no sleep, no sleep' over and over again; a major appearing at the Portaloos, smiling, noticing nothing; Male 1 using karate chops on the neck of the prisoner they called 'Grandad' because he couldn't keep his head up; punching, kicking feet, a kick to the kidneys, a fifth soldier entering one of the rooms and slapping one of the detainees about the head and leaving again, 'Get your head up or I'll slit your throat' from Male 1, a detainee lifting his sandbag for a moment, Male 1 catching him 'You thought I didn't see that didn't you?', a kick to the groin, the detainee doubling over in pain, the 'choir' singing out with grunts or groans when Male 1 went behind them and kicked them in the lower part of their backs, the guards laughing. Another soldier coming in, a big man, six feet tall, muscular, broad, bald-headed, wearing a purple T-shirt and combat trousers, entering casually, having a few words with the other soldiers and then walking up to one of the detainees, slapping him once about the head, throwing a kick at the man's kidneys so that he keeled over on to the floor, pulling the man up again and then leaving. It was shocking not just for its violence but the sheer speed and flippancy with which the soldier appeared, struck and was gone. It was all so natural, so commonplace.

SAC Hughes only left the building when LCpl Riley called out to him. The *GMTV* crew were returning. He had to go. Nothing was said.

Hughes was willing now to identify the soldiers he had seen.

WO Spence told Hughes he would have to leave Iraq, although he was warned he might have to return to make the identifications. Hughes wasn't sorry about going home, even if he had enjoyed his duties driving the press about Basra. The visit to Battle Group Main had ruined everything, he said. He was glad to get away.

Three days later he was on an army transport plane back to the UK. He was to enjoy an extended period of R & R.

25

DAOUD MOUSA WAITED FOR NEWS about the release of his son's body. The family needed to mourn Baha's death properly and to complete the funeral rituals. He had been told that this would only be possible once the post-mortem had been completed. There was a delay because a specialist was being flown in from the UK. On 20 September an SIB-appointed interpreter telephoned Mr Mousa. He was told that the autopsy would occur early the following day and once that was over he could make arrangements for the body to be collected. And yes, an Iraqi doctor would also be present, an independent witness to the procedure. Daoud Mousa, of course, was welcome to attend, at least to meet and speak with the British pathologist. He and his family prepared themselves.

At dawn the next day, Daoud Mousa and Baha's surviving brother, Ala, travelled to Shaibah base. They had told other members of the family what was happening and they too made the journey to wait outside the gates. Everyone wanted to know what the outcome of the post-mortem would be. They wanted to know how Baha had died.

Daoud and his remaining son were allowed into the base whilst everyone else was kept outside. The two of them were shown into a room where they were introduced to Dr Ian Hill and an Iraqi doctor. They spoke for a few minutes about Baha's condition before his arrest and then the medics left to prepare for the autopsy.

Hours passed. A crowd of about thirty or forty family and friends of Baha had by now congregated about the base's entrance. They were chanting and waving flags, 'causing an obstruction', the SIB log recorded. SSgt Cooper collected Daoud Mousa and took him down to the gate. She wanted to quiet the crowd, to defuse the tension, avoid any possibility of confrontation. That would interfere with everything the SIB was trying to do. She explained to the gathered crowd that autopsies took time, the pathologist had to be meticulous and they had to be patient. Mr Mousa would be allowed to see the doctors afterwards and they would learn about the findings as soon as they were available. It was a promise that assuaged any growing ill feeling and the crowd began to dissolve, people drifting away. Daoud Mousa waited.

Finally, Dr Hill emerged from the tent, quickly scrubbed down and after a pause of several minutes sat with Daoud Mousa and Ala. It was not an easy discussion. With staff sergeants Sherrie Cooper and Daren Jay and Captain Nugent present as well, Dr Hill was brutally frank about his discoveries. He said that Baha had died of ligature strangulation. Everyone heard the words. Everyone guessed what it meant. Strangulation was a murderous term not indicative of accident. It was direct and cruel, and it was also clear.

The Iraqi doctor had already spoken to Daoud about the other injuries, the broken ribs and the bruises. This last piece of medical information emphasised the scale of violence that was done to his son. It shook Daoud. To hear the details that could fuel the imaginings of suffering was a little too much. He became distraught. He demanded an official apology, compensation. How would the family cope without Baha, how would Baha's sons cope?

Members of the RMP talked to him, asked for permission to take samples, which he gave, outlined how the investigation was to proceed, tried to calm the rage. Only after some time could the SIB officers end the meeting and escort Mousa off the base. SSgt Cooper accompanied him to the gates and told him to come back the following day so that he could collect his son's body and receive the death certificate. It was a tawdry, unceremonious moment.

On the following day, 22 September, a full week after Baha Mousa had been killed, Daoud returned to Shaibah base with a car and a coffin and a local TV crew. He didn't believe that his son's death should go unnoticed. If he could publicise what had happened to his son then maybe some justice could be done.

SSgt Cooper met Mousa at the gates again. She put conditions on his entry: car, coffin, OK; TV crew, no. He didn't object. Cooper took him to the mortuary where Baha was lying in a military coffin. Daoud Mousa was asked to confirm his son's identity for one final time and then the body was lifted out and placed into the family's coffin. It was at this point that the mortuary attendant, Corporal Edwards, handed over the death certificate that had been prepared.

The certificate was rudely filled out by hand in English. It was entitled

'Death Certificate – International' and it was a one-page form to be filled in. Some of the sections showed that the certificate was intended for UK forces personnel serving overseas, as they referred to 'Rank/Rating' and 'Regt or Corps/RAF Command' and to 'Ship/Unit/RAF station'. Nonetheless, the majority of the short document could apply to anyone. The main section dealt with the 'Cause of Death' and provided a sequence of seven boxes to be completed. They referred to the 'disease or condition directly leading to death' and any 'antecedent causes', 'morbid conditions' and 'other significant conditions contributing to the death'. The form handed to Daoud Mousa included only brief entries. Even the most rudimentary knowledge of English would have shown that only two words were recorded for the cause of death.

Rather than take receipt, however, Mousa asked for these words to be translated there and then. He wouldn't be satisfied until that was done. Perhaps the brevity of entries on the form induced some prescient concern. Whatever the reason, an interpreter was called and began to translate the form. Things went awry immediately. Baha's name was incorrect. The surname was recorded as 'Dawood' instead of 'Mousa'. Then the interpreter translated the disease or condition directly leading to death. 'Cardiorespiratory arrest', it said – heart failure, in other words. And underneath, where the box required an entry noting how that condition had come about, there was hastily written 'Unknown – refer to coroner.'

Heart failure? Unknown causes?

Daoud Mousa was enraged. He'd sat in a room the day before, enduring the torture of the British pathologist describing a litany of injury inflicted upon his son, concluding that death had been the result of asphyxiation, clearly saying the way in which this might have happened. Now here was a form saying that heart failure due to 'unknown' factors was being registered as the cause of death.

Moments of deep confusion followed. The mortuary attendant couldn't do anything about the form and nor could SSgt Cooper. It was suggested that Corporal Edwards should see if he could get the form changed. It had been signed by Captain Andrew Le Feuvre, the senior house officer in the A & E department of Shaibah hospital, and Edwards went to find him. He returned several minutes later with a new certificate. The name had been changed to record Baha Mousa's proper appellation, but the captain had refused to alter anything else on the form. The entries describing the cause of death and how it came about were left untouched.

SSgt Cooper could sense the situation swerving out of her control. She knew Daoud Mousa was right about the pathologist's findings. She had been in the room at the time when Dr Hill had so bluntly relayed his opinion. There was no doubting that the form was wrong. It gave a completely false description. She had heard the term 'asphyxia' used and she couldn't deny that now. Nor did she want to. But if the senior medical officer was not prepared to change the form, who else would do so? Dr Hill had already left the country.

Daoud Mousa was becoming frantic. The tension in the room threatened to spill over to the crowd outside. Then the hospital sergeant major, Warrant Officer Pooley, stepped in. He was insistent that these Iraqi civilians, Daoud Mousa and his son and their car and their body, had to leave the base immediately. He was worried that matters would escalate and the excitable crowd at the gates would erupt. No one felt easy when a mob could form at any moment and incite a sudden upsurge in violence.

With an eye to the dangers as well as the justice in Daoud Mousa's outrage, SSgt Sherrie Cooper decided to ignore the protocol of army authority and grabbed the certificate from Corporal Edwards. She took a pen and added the word 'Asphyxia' after 'cardiorespiratory arrest', signed the amendment and handed the form back to Daoud Mousa. It was sufficient to resolve the burgeoning crisis. Mousa accepted the amended form and drove away with the body of his son, leaving SSgt Cooper to confess what she had done to Captain Nugent and record her actions in the case file. Except that the file entry wasn't entirely accurate. It said:

> Death certificate handed to father however after translating it he was unhappy with it as the name of the deceased was incorrect name should be Baha Dawood Salim Musa, he was unhappy with cause of death as it said cardiorespiratory arrest – refer to coroner. Death certificate changed except cause of death. This was explained to the father who eventually accepted it.

The reticence to record what had actually happened was understandable. But why the death certificate had made no reference to Dr Hill's findings in the first place was never satisfactorily explained. Years later, Dr Andrew

Le Feuvre would say that he couldn't remember anything about the document and couldn't even understand why he would have signed it at all. Normally he would only have done so if he had been the deceased's usual doctor or perhaps had been present at the person's death. He had been neither. His hospital records, however, indicated that he'd received a call from Dr Keilloh of Battle Group Main on the night Baha Mousa had died. At 11.54 pm, and nearly two hours after Mousa had been pronounced dead, Dr Le Feuvre had spoken with Dr Keilloh who had informed him about the death. The note he'd taken echoed Keilloh's own initial belief that the deceased had died of a heart attack. But it had also recorded that a post-mortem might be required. Dr Le Feuvre surmised that he'd been asked to write the death certificate at that point and therefore the only cause of death he could have inserted would have been Dr Keilloh's diagnosis.

Even with such a possible explanation, grounded not in memory (which wholly eluded Dr Le Feuvre), but in conjecture, this was an odd thing to do. To complete a death certificate without even seeing a body was highly irregular irrespective of whether or not Le Feuvre was under pressure at the A & E unit of Shaibah field hospital at the time. Regard for proper procedure was not *that* lax, as demonstrated by his refusal to amend the certificate (apart from the name) when requested to do so by Corporal Edwards. It was a material issue, however. It could make the difference between Baha Mousa's death being treated as manslaughter rather than murder.

26

AFTER DIFFUSING THE TENSION AT Shaibah hospital, SSgt Cooper had herself escorted to Battle Group Main along with one of her RMP colleagues, Sergeant Stanford. She had an appointment with Lt Col Mendonça whilst Stanford was to undertake the overdue crime scene examination at the detention block. Sergeant Stanford took a video of the now empty rooms and collected several sandbags, one of which appeared to be splattered with blood.

SSgt Cooper's task was to brief the CO of 1QLR and update him on all recent developments in the investigation. They sat in his office as she relayed the increasingly perilous information. Perilous, that was, for the regiment's and Mendonça's reputations. Following the post-mortem, she told him, it was clear that whatever the death certificate had recorded initially, the cause of death was *not* natural. That meant that they had to contemplate a range of possible offences: murder, manslaughter, assault, ill-treatment, abuse and possibly neglect. The surviving detainees were now being interviewed systematically and their statements were being taken. She handed him the arrest records for Private Cooper and LCpl Redfearn and explained that for the moment these men were not to be charged. But with the information they had, she said, the SIB squad would now need to move on to the battalion, interview all members of Anzio Company, all troops responsible for guarding the detainees, anyone who they thought might be able to reveal the circumstances of the death and the way in which the group of detainees had been picked up and handled. She also told him that the family were demanding an apology and compensation. Neither of these was a matter for the SIB. Both were Battalion's and therefore Mendonça's responsibility, at least to begin with. Her recommendation, though, was to get advice from Brigade before he did anything.

Later, SSgt Cooper would describe Mendonça as 'devastated' by all the news she had delivered. She would say he told her he was worried that all the good work that 1QLR had completed in Basra would now be forgotten, overshadowed. The case log also recorded that he was 'fairly angry with his regiment' and that he offered whatever assistance he could to the investigation. SSgt Cooper asked for more armed escorts. Without personnel in accompanying vehicles the RMP couldn't operate speedily, she told him. The new regulations that required all vehicles to move in pairs for their own security were preventing the investigators getting to the locations around Basra so as to finish their inquiries. Lt Col Mendonça agreed to help.

With the unpleasant briefing over, Cooper returned to the investigation. She made use of her time at Battle Group Main by re-interviewing some of the medics who had treated Baha Mousa on the night of his death. She had already questioned them on the 17th and had taken short statements then on the details of their treatment of the dead man. This time she wanted to know whether they may have noticed or treated injuries to other detainees. She went over to the Regimental Aid Post and spoke to the three medics there. By some weird coincidence two of them had precisely the same name, Steven Winstanley, although one was a corporal and the other a private. The third medic was Corporal Baxter.

Private Winstanley told SSgt Cooper that he, his namesake and Baxter had been assigned to examine the detainees on the day of their arrest, 14

September. He'd visited the detention facility at about 3 pm under the supervision of his two colleagues and had seen eight prisoners, standing up, hands cuffed together, arms lifted up and out towards the front at shoulder height, their heads covered with sandbags. A brief check of each man had been conducted, the sandbags lifted so that their faces and necks could be inspected, shirts too to see their torsos. The medic hadn't seen any injury or physical problem at all, they had all appeared fine to him. He hadn't bothered to fill in any form, he said. There was no need, nothing to record.

Cooper returned to HQ, her case hardly advanced by seeing the medical team.

27

EVEN IF THE MEDICS HADN'T seen anything, there was now credible evidence that the detainees had been subjected to a long period of abuse at the hands of Anzio Company members and other soldiers at Battle Group Main. But allegations of assault or worse would need careful substantiation. The detainees would have to be witnesses. It was imperative that their stories were sufficiently robust to support a possible prosecution. This was by no means a certainty. Hadn't they been hooded for much of the time when in custody? Wouldn't this impair their testimony?

WO Spence was detailed to take an SIB contingent to Camp Bucca to record more complete statements from the prisoners still held there. On 23 September he and his team began the process. Interviewing was tortuous. There was a lack of qualified interpreters so progress was bound to be slow. But gradually, a more complete picture of what had happened during those thirty-six hours between 14 and 16 September took shape. Seven prisoners were seen.*

---

* Only three detainees in addition to Baha Mousa can now be named without incurring possible contempt of court proceedings. One, Kifah al-Matairi, died in 2006. He was killed in a bizarre accident when the roof of a temporary house he was living in collapsed on top of him. An iron beam fell on his neck and killed him instantly. Ahmad Taha Musa al-Matairi, Kifah's brother, and Ahmed Maitham chose not to remain anonymous when the judicial inquiry into the death of Baha Mousa began in 2009. The others wished to avoid any publicity and they were given labels from Detainee 001 to Detainee 006. It was a device to preserve their privacy.

The statements collected from the detainees that day were shockingly powerful. They indicated that not one but many crimes had been committed.

The hotel generator operator, like Baha Mousa, was a relative newcomer to his job, taken on only about five weeks before the arrest. During the raid he saw one soldier push his booted foot on to Baha's head, one, two, maybe three times as he lay prone on the floor of the reception area. The soldier didn't apply his full weight as far as he could see but it was enough for Baha to cry out each time. He remembered they were taken to the British camp, where two sandbags were put over his head and he was forced to bend his knees and put his arms out in front of him and maintain that position. Water was given to him if he asked for it, but sometimes it would be poured over the hoods or squeezed through the coarse fabric so that he couldn't drink properly. He was kicked and punched in what he called 'the beating room', but he couldn't see who was responsible as the sandbags were kept on during the attacks. He thought he was going to die, the blows were so vicious. They concentrated on his kidneys, about forty kicks, he thought, all in the same region. He wept and screamed and eventually collapsed. Some soldiers took him to a medical centre where he was allowed to rest for a few hours. Afterwards he was taken back to the facility, to a different room, not the 'beating room'. They let him sleep there but the rest didn't last. The next day, or so he thought – the ordeal confused his sense of time – four soldiers came into the room, picked up the plastic drinking bottle he had been given to urinate in during the night, and poured the piss over his head and into his mouth. They laughed and then took him back to the 'beating room'. Eight to ten soldiers were waiting there. One was shouting, shouting, and several kicked the prisoners. Another was taking photographs. He remembered that one posed beside him with his fist clenched by his head, a photograph was taken and the soldier punched him. He was made to dance. 'Come on Michael Jackson' they shouted at him, and he was forced to move his legs and arms up and down. On the second night he heard Baha Mousa cry out in Arabic 'Mercy' and 'I'm dying.' The voice was coming from a different room within the building. That was as much as he could remember.

The experience of the hotel cleaner and part-time guard varied slightly, although many of the details matched the generator man's account. His shift at the hotel was between 6 pm and 9 am. If the soldiers had come two hours later he would have missed them, but he was not blessed that day. It all became horrific when the soldiers found guns in the reception area. He was made to lie on the floor with the other hotel employees. After a while he was taken to the toilets where he was made to sit over the hole whilst the soldier pulled the chain, splashing the soiled water over his trousers. He remembered that a hood was placed over his head before he was taken to the trucks outside the hotel waiting to take them to the British Army camp. When they arrived at the camp they were taken inside a smallish concrete building where he was searched. His hands were tied to the front and he was forced into the bended-knee position, the stress position. He became tired and experienced cramp and tried to straighten his arms and legs. Soldiers kicked and punched him when he did so. They aimed for his lower back and his kidneys. Throughout the first day he was hit about the head. It felt like they were using a sandal or a shoe, he said. He wasn't fed and when he was given water it was tepid, hot even. Sometimes they would spray it into his mouth so that he couldn't swallow and it would squirt out of his nose. A metal bar was struck against the floor to keep them from falling asleep. That was the first day. On the second, they gave him food, beans, something he described as 'dumplings', bread. It was all cold. The kicks and punches continued and one time someone tried to push their fingers into his eyes through the hood. He could hear the cries of others around him, cries from his compatriots from the hotel.

The hotel's main security guard had worked there for about six months. On the day of the arrest he was just about to finish his night shift. Then the soldiers came. He said he was made to lie flat on the floor and a soldier stood on the small of his back. After a while they were taken to the British base where he was moved to a small building, made to endure the stress positions with the other detainees, the yelling, blows to his back and face, hard, compact, deeply painful. They all screamed, he said, and wept intermittently as they were targeted one after the other. Finally he was allowed to sit on the floor, but the kicks and punches didn't stop. A metal bar was banged against the concrete to stop them sleeping. One soldier put his arm around his throat and squeezed. He thought he would suffocate. Later, someone sat on his shoulder and farted. The guards laughed at that. They were having a good time. The shouting carried on and on. He didn't understand English so he couldn't say what the soldiers were trying to tell him. Once he was punched in the eye, but he didn't fall over as he feared he'd be kicked to death if he ended up on the floor. At some point during the second night he saw Baha taken away on a stretcher. Then he was told they were allowed to sleep. Breakfast arrived the next day, some yellow stuff he couldn't identify, with spicy beans.

Ahmad Taha Musa al-Matairi was part owner and manager of the Hotel al-Haitham. On the day of the raid on his hotel he was at home, asleep in bed. He received a telephone call from his brother Kifah telling him soldiers were at the hotel and were searching the place. They needed the key to the safe. He dressed quickly and took a taxi to the hotel. He saw the army vehicles outside when he arrived and on entering the reception he noticed members of his staff seated on the ground with their hands on their heads. He handed over the keys and was immediately pushed on to the floor with the others. They were all made to lie on their stomachs. He saw some rifles and pistols put on the front desk. Al-Matairi said these were for the hotel's protection, although he was sure the pistols were owned by his partner in the hotel. The hotel guests were brought down to the lobby and searched, but they weren't made to lie down. Then, he said, the staff and he were all taken to the hotel toilets for an hour or so, where the soldiers pulled the chains so that the Iraqis' clothes would be splashed with the dirty water from the latrines. Finally they were loaded on to trucks and driven to the British base.

Once in the detention facility, al-Matairi was treated as all the other men. He remembered saying to the soldiers, in English, that he was an educated man. He pleaded with them.

'Why are you doing this?' he said. 'When the British arrived in Basra I took my children into the streets with flowers to welcome you.'

But he was told to shut up. The beatings rained down on him, so heavy sometimes that he couldn't breathe. He was kicked in the groin and collapsed on to the floor. He was already suffering from a small but visible hernia and the lump began to grow. He said he complained about feeling very sick and was taken to a medical room where a doctor examined him. His hood was removed during the examination but put back on when he was returned to the detention building. He remembered later hearing Baha crying out in the night 'I'm going to die.' He had known him for ten years and recognised his voice. Baha shouted 'I do not support Saddam, I'm not Ba'athist. Why do you do this?' Then he heard nothing more. The next day food arrived and they were put in trucks and taken away. He had an enlarged hernia and severe bruising to his kidneys and back.

Mr al-Matairi's partner in the hotel wasn't there on that Sunday morning. He was at home. Unknown to him, the soldiers who had raided the Hotel al-Haitham had found out where he lived and had gone there seeking the owner. He was rousted from his bed and forced to lie on the ground outside his house. He had a heart condition so was too scared to challenge what was happening. After a time he said he and his son were taken to a British Army base together. They were kept there for a few hours and then transferred to Battle Group Main where at first they were put into a room on their own. At least, so he thought. He was wearing sandbags over his head by this time and couldn't be absolutely sure no one else was with them. But he heard groans and screams coming from another part of the building and remembered that he too was forced to maintain a stress position and was beaten if he let his arms drop. He was hit over the head with a torch, he said. If he started to fall asleep at night a soldier would call out 'No sleep.' Then the next morning he thought he was having a heart attack, the pain in his chest was so intense. He lost consciousness for a while. When he came to, his hood and plasticuffs had been removed and there was a soldier standing over him with a stethoscope. The man was very kind to him, he said. He was given pills. From that moment he was left alone and not mistreated, but he could hear the shouts of others from else-where in the small building and he witnessed a number of soldiers beat a prisoner who shared his room.

It was the voice of his mother screaming which woke the teenager, the eighteen-year-old son of Mr al-Matairi's partner, that Sunday morning of 14 September. He was in bed in his father's house. His door was pushed open and three British soldiers carrying rifles came in. They took him out into the corridor, past his mother and sister who were on the floor shaking with fright. The house was in turmoil, the sound of doors cracking open and shouting coming from the various rooms and floors. He was taken outside where his father was already kneeling on the floor with his hands on his head and a soldier holding a rifle standing behind him. They were both kept in this position for some time until they were eventually thrown into an army truck and driven a short distance to a British base (not Battle Group Main). Here, the two men were told to get out of the truck. The boy said he was told to take his shoes off and was then made to jump up and down for what seemed to him to last an hour until he collapsed. Water was poured over his face and he wasn't allowed to use the toilet when he asked. After a couple of hours he and his father were pushed into separate Land Rovers. He was kicked in the spine. The vehicles were driven a short distance again, this time to Battle Group Main, where his father and he were taken out of the Land Rovers and into the detention facility where all the other detainees were housed. Plasticuffs were used, sandbags too. He said someone told him in broken Arabic 'no water'. He remembered that one of the guards pissed on his head. He couldn't see who it was. At some point he was taken outside, he could feel hot sand under his feet, and he was pushed down to sit next to a generator. His head was still covered by a sandbag. The generator's exhaust was searing, but he was pushed close to the machine and made to stay there. He was kicked and punched and began to cry. A soldier told him to shut up. 'No noise', he was told. Then, around midnight (one of the soldiers told him the time otherwise he wouldn't have known) he was taken to a different room elsewhere on the base, and asked questions. The officer who questioned him treated him well, gave him water. There was an Iraqi interpreter there too and at one point, when the boy said he had asthma, he heard the interpreter tell the British soldier that this was a lie. The boy said he was then returned to the detention block, but this time he was pushed into the middle room, the toilet, where he was made to kneel in the foot grooves either side of the old latrine with his wrists tied behind his back, his head bent towards the hole, which still stank of excrement. The cries of other men, screams, grunts, shouts, continued throughout the night. He recognised his father's voice amongst them. He was kept kneeling over the toilet for a few hours until eventually he was taken back for interrogation, questioned again, then returned to the same toilet. In the morning of the next day he was allowed back into one of the main rooms. They seated him next to his father. A soldier seemed to take pity on them. Food was brought, but they didn't eat it.

They were both too scared. Much later, that night, he saw the hotel receptionist, Baha Mousa, being carried out from the toilet where he had spent so much time the night before. He didn't see Baha again. The sympathetic soldier returned and untied their hands. Food was brought which again they didn't eat. During the night they were kept awake. They were told 'no sleeping', and soldiers would shout at him 'I'm going to fuck your mother', and they asked how much it was to sleep with an Iraqi woman, but he didn't answer even though he could speak some broken English and understood what they were saying. The next day, they were told that they were to be taken to another camp. One soldier said they would have their heads cut off there, drawing his finger across his throat. He was terrified.

There was one more detainee who had *not* been arrested in the Hotel al-Haitham on that Sunday and had absolutely no connection with the arrests made there. His name was Ahmed Maitham and he only appeared in the Battle Group Main detention facility later on in the day, sometime after 9.30 pm. His arrest followed a traffic accident in Basra. British troops had come to the scene of the crash and found three AK-47s in the boot of his car. Initially, he'd been taken to the Iraqi police station at Hay al-Hussein. He'd been treated well there, but then had been loaded into a Land Rover and driven away. During the journey he was kicked by a couple of the soldiers in the back of the vehicle. It wasn't hard, but it was malicious. Eventually they arrived at the British base where, like all the hotel detainees, he was taken into the detention block. It was night and inside it was dark, but he could make out five or six Iraqi men sitting cross-legged with sandbags over their heads and their hands cuffed in front of them. He was moved to the other room in the block, though, where he was told to keep his arms stretched out in front of him. If they fell he would be kicked. This happened many times. His shoulders hurt, his back hurt, his kidneys hurt. He endured this treatment until finally the soldiers put him into a truck with other detainees and drove them away.

These then were the seven accounts that Spence and his colleagues in the RMP recorded at Camp Bucca. They were a little terse, a little lacking in narrative clarity, but taken together they provided a powerful vision of two days of madness and suffering.

But were the detainees lying? Had they secretly huddled together and fabricated their accounts of ill-treatment? Since their transfer to Camp Bucca they'd been kept in the same tent in the British compound at least for a time. What was to stop them making up those stories about the minor assaults in the hotel and then the more serious episodes in Battle Group Main detention facility? Couldn't they have concocted those details about being urinated on, having guards break wind over their heads, made to dance like Michael Jackson, having tepid water squirted through their sandbag hoods, kept awake at night with an iron bar, treated with disdain and humiliated and beaten on countless occasions during their ordeal? That would have served their purposes, wouldn't it, if they really were Iraqis still loyal to the former regime, or simply opposed to the invasion by the British? If believed and publicised, these stories about the misconduct of so many British soldiers would be a great propaganda coup for those fighting the Coalition.

The theory that some collusion had occurred so as to construct a story discrediting the men of 1QLR and therefore the whole British presence in Basra, seemed improbable to WO Spence. Of course, some of the accounts didn't match exactly. The owner of the hotel and his teenage son were kept together most of the time, taken to a different camp from the other detainees before being moved into the detention building at Battle Group Main, and yet their stories varied in some important details. The father, for instance, said nothing about his son being forced to jump up and down at the first base where they had been taken. But the minor discrepancies which appeared were understandable given that hoods were used consistently, that the prisoners' individual experiences were bound to be different with the number of soldiers encountered and the prolonged period in which assaults of different kinds allegedly occurred. Besides, the SIB had already seen the evidence of injuries sustained by the detainees and these supported the stories of beatings. They had the evidence of Riley, Hughes and Betteridge, whose independent information matched the accounts the prisoners had given. And they had the dead body of Baha Mousa sporting a plethora of wounds which a leading British pathologist of unquestionable professional character had confirmed would have been caused by sustained and vicious beating.

With such interlinking and corroborating evidence it was simply not credible that the detainees had fabricated their testimony.

WO Spence couldn't have doubted that he faced an investigation that threatened the reputation not only of the regiment, but the whole British Army. That would have daunted the most impervious detective.

28

KEPT IN CELLS WITH TEMPERATURES *over forty degrees centigrade; restricted visits to toilets and left to soil themselves; prevented from sleeping; forced to run in circles; subjected to psychological torture and threats; systematic beating; humiliation and restriction of food; all part of an interrogation routine designed to extract information from suspected terrorists. The British Army had been here before. In 1966 the Swedish section of Amnesty International published a report on the alleged inhuman treatment of detainees in Ras Morbut interrogation centre, Aden, now called Yemen. The official response from the British government was outrage that such accusations could be levelled at British troops. But when investigated it became clear that British forces attempting to acquire information about the uprising in the territory cared little for the Geneva Conventions or human-rights niceties. There was slight, if any, moral control.*

The experiences in Aden and Cyprus and Northern Ireland and other conflicts in ex-colonies made the British Army and its political masters acutely aware of the possibility of prisoner abuse. The Ministry of Defence published a manual of the law of armed conflict (the latest version with Oxford University Press in 2005) as a reminder of the legal commitments. Much of its content was common knowledge at the time of the Iraq War in 2003 since the main treaties and conventions were already in place and well known by army lawyers, ministry officials, and serving officers and men. At least, their basic principles should have been familiar. There was nothing obscure about the contents of the Geneva Conventions.

Indeed, few if any soldiers in the modern British Army could argue persuasively that they didn't know what the Geneva Conventions were. They may struggle over some of the detail, and the current Ministry of Defence manual is over 500 pages long and covers a bewildering range of obligations and rules, but the fundamentals are hardly unintelligible.

They had been forged over centuries, inspired by different notions of chivalry, honour, humanity and mercy, by Christian, Hindu, Islamic, Sikh religious teachings. Even in medieval times they were well known. In Henry V,

Shakespeare's Fluellen cried that the king's decision to kill all the prisoners on the Battle of Agincourt was 'expressly against the law of arms'. The modern-day soldiers' codes of conduct, however, have their genesis in mid-nineteenth-century Europe. A Swiss travelling businessman named Jean Henri Durant, a man with an irrepressible appetite for work and an active social conscience, happened to be at Solferino, a small town in the disputed state of Lombardy in northern Italy, on 24 June 1859 just as the battle of that name between the French Army of Napoleon III and the Austrian forces ended. It was a French victory, as if he cared, but it was the aftermath that haunted M. Durant. He meandered about the battlefield strewn with corpses and the shredded flesh of still living men and vowed to do something to relieve the appalling suffering of those required to fight. With evangelical enthusiasm he returned to Switzerland to found the International Committee of the Red Cross, an organisation devoted to protecting the lives and dignity of the victims of armed conflict. The core task was the construction of an international law designed to change military practice and reduce suffering.

It took time, but gradually people became attracted by the idea that the way war was fought could be regulated, that soldiers could be trained to behave so as to reduce the level of harm done to civilians and even the enemy in some circumstances. Codes were drafted. Countries were encouraged to sign them and teach their armies to abide by them. Punishments for breaches of the codes were devised and the idea of 'war crimes' became embedded in popular culture, in law, and political rhetoric. The Geneva Conventions, the codes initiated by M. Durant after his disgust at the sight of the Solferino slaughtered and suffering soldiers, became shorthand for the multiple obligations armies were now obliged to respect.

Despite the humanitarian pedigree, there was, and still is, more than a slight odour of absurdity about all of this, though. The laws of war (as they are now often conveniently called) instruct soldiers to avoid inflicting 'unnecessary suffering' either through the weapons or tactics they use. They ban certain arms and armaments like crossbows or dumdum bullets (which explode inside the body, eviscerating the target) or mustard gas. They have much less to say, though, about Uzi sub-machine guns that can fire 600 rounds per minute and rip a person to shreds in seconds. Or artillery shells that burst on impact and emit shrapnel which can tear limbs and flesh from the body. Or banks of missiles fired in unison from helicopters to create a wall of fire and explosion. Or the innocently named 'daisy cutter' bomb, which when detonated is so powerful

that it creates a shockwave and an air burst that lays waste to a vast area of
ground. Or flame-throwers, or mortars, or rocket-propelled grenades, or bayonets,
or shotguns, single shot or pump-action, or club-like knives that can disembowel
with one practised slash. The weaponry that kills and maims is seemingly limitless. So what sense is there in banning some and not others? Is the suffering
inflicted any less?

One area of the laws of war where there is less moral ambiguity, however,
is in the treatment of prisoners of war. People hors de combat or those made
prisoner must be protected and treated humanely. Some such obligation was
the expectation even at Agincourt in 1415, as Shakespeare recorded. Civilians
in particular are entitled to respect: for their lives and for their dignity. They
must be guarded against all acts of violence and brutality. No one, the
Conventions say, shall be subjected to physical or mental torture or cruel or
degrading treatment. It couldn't be clearer. Prisoners of war and detainees can
be asked questions, but there can be no coercion through violence. This isn't
an optional requirement.

Although notorious figures in the USA administration sought to quibble
about the obligation in the wake of the 11 September 2001 attacks against the
World Trade Center in New York, few politicians accepted publicly that a 'war
against terror', as it was called, justified ignoring these standards. Whether
people believed that any ill-treatment was excusable where suspected terrorists
were concerned is another matter. There are some who argue that the 'ticking
bomb' scenario excuses torture. This is the fictional situation loved by philoso-
phers where a terrorist knows the location of a nuclear bomb primed to explode.
What should those who have captured him do? If torture would reveal the
bomb's position, would it be right to inflict it on a suspect? Saving perhaps
millions of lives for the sake of one person's suffering is often portrayed as a
reasonable deal.

The Conventions, though, are steadfast. There are no exceptions to the ban
on torture. Nor does the Ministry of Defence manual recoil from the absolute
nature of the obligations. It says that violence to civilians in captivity, torture
of all kinds, corporal punishment, outrages against personal dignity, any
form of indecent assault, are prohibited. It doesn't matter what the benefit of
torture might be. That's not an equation granted any validity in the laws
of war. The manual also acknowledges the need to educate the armed forces
about these absolute rules. The British government even signed and ratified
the Rome Statute of the International Criminal Court, which said that if

British personnel committed serious breaches of the Geneva Conventions then they could be prosecuted as war criminals. It also reinforced the concept of command responsibility. If senior officers, politicians even, failed to exercise proper control over their men, knowing or suspecting that abuses of prisoners were taking place, then they too could be prosecuted.

But how much distance lay between this official posturing and the practices of the armed forces? Had there been loss of 'corporate knowledge'? If so, who was responsible?

29

9 AM, 23 SEPTEMBER 2003. As if to acknowledge the wrong done to Baha Mousa, as if to accept that his death was an aberration, Lt Col Mendonça visited Daoud Mousa to offer his apologies for the killing of his son. Perhaps it was Mendonça's small act of atonement, a penitent gesture in recognition of the loss not only for the father but also Baha's children. Mendonça had young sons of his own, waiting for him in Britain.

A troop of soldiers from 1QLR descended on Daoud Mousa's house, securing the area whilst Mendonça sat with the family. It must have been a difficult mission. What could Mendonça have said as he drank bitter coffee and spoke through an interpreter? Daoud Mousa would recall later that the lieutenant colonel apologised and expressed sympathy and assured him that those responsible would be punished under British law. But there was no discussion about the intricacies of the laws of war and the liability of commanders for the behaviour of their men. Doubtless, Mendonça would have said little to assuage the grief or to explain how it was that Baha had ended up being beaten brutally for two days whilst in *his* compound and in a building within sight of *his* office balcony. And the men who used their boots and fists to leave the array of bruised imprints on the son's body were *his* men and under *his* command.

There were other visits after this first one. Mendonça would drop in to the Mousa home, making a habit of calling to sit with Baha's orphaned children, sometimes bringing them toys instead of apologies. But these visits had to end. As his tour of duty expired in October, he had to introduce Sergeant Gillingham of the Royal Military Police as the family's liaison officer. This was the man who would keep them informed of the investigation's progress. Daoud Mousa wasn't too happy about this. He thought it meant his son's death was no longer worthy of a lieutenant colonel's time. Sergeant Gillingham explained that he was closer to the investigation, that it was a positive development and that he could keep the family much more in touch with the inquiry. The lieutenant colonel had so many other duties to perform in those few days remaining before he and the rest of the battalion would be leaving Iraq.

At some point, no one can remember exactly when, Mendonça also raised the sensitive topic of money. Interim compensation of \$3,000 was to be paid immediately. It must have seemed a pathetic sum for a life, even if it was only a provisional payment. And there was a final gift, if gift is the right word. The Ministry of Defence had agreed, so Mendonça said, that in the event of a court martial of those responsible for Baha's death, the British government would arrange for Daoud to attend the hearing at Her Majesty's expense. That was the least they could do, although as it turned out, Her Majesty's Government couldn't even manage that.

Lt Col Mendonça left the Mousa home with his protective convoy of Land Rovers. He wouldn't return. But his future was to become more entwined with this Iraqi family than he could possibly have believed. He might not come to suffer the agonies of a tortured detention. He might not lose a son through the malice of others. His life was nonetheless to become defined by the death of Baha Mousa.

30

THE DAYS OF ENDLESS SERVICE in conflict zones, where men were forced to endure months and even years without seeing home, were not to be repeated in Iraq. Soldiers were granted home leave with satisfying regularity. Perhaps it kept them sane. Perhaps it took their minds off the tasks in hand as they counted down the days until their next trip home. Whatever the wisdom of these frequent absences from theatre, the turnover of personnel made life difficult for the SIB and their investigations.

After the detainees had been interviewed at Camp Bucca, the SIB officers were now ready to see a host of witnesses-cum-suspects. But SAC Hughes from media operations had gone home, privates Kenny and Aspinall were flying out immediately and LCpl Redfearn was following them in a couple of days. The detectives would have to work around the absences.

There were two people who were available, though: the 'tactical questioners' whose job had been to interrogate the detainees at Battle Group Main on 14 and 15 September. The detainees' statements hadn't included any complaint about their treatment during questioning; indeed their statements all seemed to emphasise how polite and considerate the interrogations had been. But the TQers might have something to say about the state of the prisoners when delivered to them. There were also a few issues that needed explanation: what was tactical questioning all about? What did 'conditioning' mean? Who was in charge? Why were these Iraqis kept at Battle Group Main for so long when detainees were supposed to be transferred to Camp Bucca within fourteen hours of arrest?

On 26 September SSgt Sherrie Cooper formally stepped down as officer in charge of the investigation and handed the file to WO David Spence. He was already intimately concerned with the case, but his other investigation had now finished and he was ready to take over the 1QLR matter. It wouldn't leave him for another three years.

Spence's first action in charge was to interview the TQers, Staff Sergeant Mark Davies and Sergeant Ray Smulski.

It was no secret that the detainees had all been subjected to tactical questioning. That was the reason why they had been transported to Battle Group Main. The battalion was authorised to arrest anyone suspected of terrorism or of being a threat to British forces and take them back to base for 'processing'. This was all part of standing orders. The detention facility was to house any prisoners whilst the battalion questioned them to determine whether to release them, hand them over to the Iraqi authorities (if suspected of only an ordinary crime), or pass them on for more sustained interrogation at Camp Bucca. Soldiers trained as tactical questioners were called from Brigade for the initial questioning at Battle Group Main. No one on the base had the necessary skills.

For Brigadier Bill Moore, like other senior officers, TQing wasn't simply a matter of process. He was fed up with the frustrating delay in information emerging from Camp Bucca. Quick intelligence was vital to Brigade's ground operations. If they were to strike at the insurgents, they needed to know where armed cells were hidden. Waiting for the Camp Bucca interrogators to produce anything useful was taking days, even weeks. Tactical questioning was a way to circumvent this delay. As long as there was someone available, trained to conduct effective questioning, someone with

the skills they taught at Chicksands, the home of the Defence Intelligence
and Security Centre back in Britain, then TQing could be a vital component in the work of the ground troops. Unfortunately, there weren't that
many qualified interrogators in Basra in September 2003. Only four were
working for Brigade at the time and these men had to serve all the Battle
Groups.

SSgt Mark Davies was one of these rare individuals. He'd taken the
TQing course at Chicksands in January 2003 prior to being deployed to
Iraq. The course had lasted twelve days and by the end of it he had a good
grasp of the Geneva Conventions, or so the report on his participation
confirmed. He was posted to Basra and farmed out to the various Battle
Groups whenever needed. He worked closely with the relevant Battle
Group internment review officer, men like Major Peebles at 1QLR, and
advised on process, carried out questioning at the Battle Group camps,
and organised the transit of detainees to Camp Bucca for more intense
interrogation once he'd finished with them. He wasn't responsible for the
care and custody of the detainees he questioned; that was a matter for the
provost staff.

Davies told WO Spence that he'd been called to Battle Group Main
early Sunday morning on 14 September, the day Baha Mousa and the
other detainees were arrested. He arrived at about 9.30 am and met
Major Michael Peebles, the BGIRO. His briefing left him with the
impression that 1QLR had captured several suspected insurgents, perhaps
had even uncovered an insurgent cell. There was a suggestion that they
might have been the ones responsible for the murder of the three RMP
officers on 23 August. At least that was what he was told. They had
recovered weapons, documents and a large bag of cash. It could have
been a hub for terrorist operations. And he was informed that one of
the Iraqis at the hotel, an elder son of the hotel's co-owner and himself
apparently part owner of the business, had managed to walk away before
the sweep by Anzio Company. If these were insurgents, then they needed
to establish the escaped man's whereabouts immediately. This was to be
the goal.

Worried that he might not be able to process all the detainees captured
in time to meet the usual requirement to transfer prisoners on to Camp
Bucca within fourteen hours, SSgt Davies called for help. He needed an
interpreter and another trained questioner. In his experience, once detainees

were transported to Camp Bucca there was little chance of any useful and reliable information filtering back to the Battle Groups.

WO Spence wanted to know what TQing involved. SSgt Davies began to talk about the techniques he'd been trained in as though they were some kind of esoteric knowledge. The 'shock of capture' was intoned like a secret enchantment enabling access to the truth in a defeated fighter's heart. 'Conditioning' was the method by which that state of mind, that terror felt when first taken prisoner, when expectations of impending death were heightened, would be preserved. It was in this zone of animalistic anxiety, fermented by isolation, an 'almost' ruthless regime of detention, disorientation, provision of only basic needs, which could make a prisoner talk. This was what he'd learned at Chicksands, the staff sergeant said.

It had taken Davies most of the Sunday to prepare himself. Indeed, it wasn't until about 7.15 pm in the evening of the arrest when finally he felt ready. This made little sense in terms of his training. How could he have taken advantage of the 'shock of capture' by leaving questioning so long? What had happened to the urgent need to discover the escaped hotel owner's hideout? SSgt Davies couldn't explain. He said he had been weighing up his tactics, readying himself for the task. But for ten hours? That meant that if the fourteen-hour rule was applied, he would only have had four hours before the detainees had to be transferred to Camp Bucca. With nine men to question, how did he think he could have managed that? No credible explanation was given.

Eventually, though, SSgt Davies said he had announced he was able to start. He called for the youngest of the detainees first. The eighteen-year-old younger son of the hotel's co-owner and brother of the escaped man was brought from the detention block to the interrogation room, his head covered in hessian bags, hands cuffed, and left standing in the middle of the room. The hood was pulled off and a 'harsh TQ session' began, as Davies described it, firmly in accordance with the training manuals. Three to five minutes of intense shouting an inch from the detainee's face, apparent aggression, imputed threat, but no physical contact; this was what he meant by 'harshing'. After this, Davies had the boy taken away to sweat and then brought back later in the evening to be questioned again. Davies said it produced results. The boy talked about the escaped man's connection with the Ba'ath Party and purchase of false ID papers to pass through Coalition forces' checkpoints. But he didn't reveal the fugitive's whereabouts.

SSgt Davies told Spence that he worked steadily through a couple of the detainees until a second tactical questioner, Sergeant Smulski, turned up in the early hours of Monday 15 September. Davies had watched as Smulski brought the boy back for a third interrogation between 3.30 and 4.15 am. He wanted to see if Smulski was up to the task.

Apparently satisfied, SSgt Davies ended his shift. He wrote up his notes in the early hours and departed Battle Group Main after breakfast on the 15th, leaving the remainder of the questioning to Sergeant Smulski. Davies said he had no further contact with the detainees. He swore there 'was nothing done nor did I witness anything in relation to the guards or the prisoners that gave me cause for concern'.

Had SSgt Davies told everything he knew? Nothing had been mentioned about the generator where the eighteen-year-old boy had said he'd been placed for an hour, enduring the suffocating heat of its exhaust and the ceaseless deep drone of its engine. And nothing had been said about how the detainees were 'conditioned', softened up.

WO Spence moved on to Sergeant Smulski. His interview was a little more revealing.

Smulski told Spence that the extent of his training in tactical questioning was both out of date and negligible. In 2000 he'd attended a one-week course on the subject in Staffordshire. That was all; no refresher training, no update, no further examination. A 'pass' was sufficient for him to be sent to Iraq and thrown into the shadow world of insurgency and terrorism. Smulski was nonetheless able to remember the basic principles of the course to help him. Like Davies he'd been initiated into the secrets of the 'shock of capture'. He knew prisoners should have a 'fear of the unknown' instilled in them, they should be kept awake, kept tense in that physical and mental condition when first taken into custody, in that state of anxiety designed to encourage cooperation.

When Smulski arrived at Battle Group Main, pulled from his bed by his commanding officer at 10.30 pm on 14 September and told to hurry over to help with questioning, he, again like Davies, was briefed by Major Peebles, 1QLR's internment review officer. Smulski said he was told to take over from SSgt Davies and continue to question the nine Iraqis brought in earlier in the day. Peebles showed him a room at the rear of the headquarters block in the compound. It had a curtained door, was about nine feet by twelve feet and bare except for a rough table and a bench seat. The strip lighting and white tiled floors and walls must have made the room eerily antiseptic. There was a grate in the floor suggesting it had been a shower room once, but Smulski didn't notice any taps or water source. He remembered that SSgt Davies was interrogating one of the detainees with a guard on the door, another sergeant taking notes and an interpreter standing behind the detainee translating the questions and answers. Smulski watched and learned.

After a time, Smulski wandered over to see the detention facility. He hadn't been there before and wanted to see what it was like. When he entered he noticed one of the detainees' shirts was ripped, exposing his midriff. Smulski saw bruising and he asked the guards whether the medical officer had attended to him. Yes, they said. He gave it no further thought. But he wasn't happy with the preparations for questioning. Perhaps with the thought of the 'shock of capture' in mind, he set about having the prisoners seated on the floor, facing the wall and kept apart from each other. He was adamant that they shouldn't be allowed to sleep. Make as much noise as possible, he told the guards. Keep them awake. He instructed the guards on how the detainees should be brought to his interrogation sessions: the prisoners should have their hands cuffed before them, he told them, they should be picked up from the floor by hauling on the clothes at their shoulders, and they should be led across the compound to the questioning room by both thumbs in one hand. That's what he'd been taught on his course. It was 1.40 am.

Throughout the night Sergeant Smulski conducted interrogations. Like SSgt Davies, he started with the eighteen-year-old boy, whose youth only registered as a vulnerability that could be exploited. Back in Britain, it might have triggered alarms, but this was Iraq and the prisoner was an Iraqi 'male'. There was no concept of the interests of the child being paramount here.

Smulski revealed few other details about the questioning to WO Spence, but he remembered being told by the provost corporal (undoubtedly Corporal Payne) that one of the prisoners was giving them trouble, slipping out of his cuffs and lifting his hoods. The man was being kept apart in the middle room of the detention facility, the old toilet. Smulski wanted to be helpful and told Payne that the detainee should be laid face down on the ground, arms forward under his head. They went to see the prisoner and Smulski said he and Payne forced the man into position to show how it could be done. He didn't mention that the room was dominated by a hole in the ground, the one which stank of excrement and the one next to which the awkward prisoner was forced to lie.

Sergeant Smulski finished his questioning at 4 pm on 15 September. Then he left, he told Spence, and had nothing further to do with the detainees.

31

THE DESIRE FOR HUMAN INTELLIGENCE, 'HUMINT' in army circles, was the insidious thread which bound Baha Mousa to the detainees from the Hotel al-Haitham, to the detention facility in Battle Group Main, to the treatment of those Iraqis taken prisoner, to Camp Bucca, to the field operations of 1QLR, to the tactical actions carried out by the British Army in Basra, to the intervention in Iraq by Coalition forces, to the invasion of Afghanistan, and, ultimately, to the 9/11 suicide plane attacks in the USA. For, perhaps more than anything else, the destruction of the World Trade Center in 2001 was perceived as a failure of intelligence. Either no one knew a threat of that magnitude existed or no one took it seriously. The intelligence 'community' had failed. Their excuses were long-winded, but fairly simple at heart: they didn't have sufficient networks of informants and reliable analysts capable of penetrating fundamentalist Islamic groups. They didn't understand their culture, knew little about what motivated them, had no 'feel' for the violence they might unleash. That suicide on such a scale should figure so largely in their plans was truly shocking. Not that there hadn't been warnings. Other terrible atrocities and suicide bombings had happened before.

The absence of intelligence prior to 9/11 haunted governments. It might be thought impossible to stop the determined terrorist from blowing himself up in some crowded public place, but the extraordinary coordinated hijacking and crashing of four airliners should surely have been picked up on the intelligence radar. Fear of a gaping hole in information gripped security departments across the western world. The terrible question was: if we don't know about this, what else aren't we aware of? And if 'they' are prepared to sacrifice themselves and others on such a scale, what else will they do? Nuclear attack?

Chemical weapons? Biological warfare? The USA government understood the extent of the possible peril because they had access to all these threats, stockpiles of all three. They might have considered themselves responsible enough not to use them, but if these weapons existed 'out there' would the extremists be so constrained? No, was the operating assumption. But if it wasn't known who 'they' were, or where they lived, or how they operated, or what they believed, or even what they wanted, how could any protection be effected?

The only solution was attack, to declare war. And as with any conventional war, intelligence was vital to ensure success in the coming fight.

Basra in September 2003 was enmeshed in this logic. Coalition forces were sent into Iraq prepared to uncover information. That was what Camp Bucca was all about. That was why the British Joint Forward Intelligence Team was there. That was the rationale for interrogating anyone suspected of posing a threat to Coalition security. The same desire for information about an opponent obscure to sight and understanding drove the need to soften suspects, subject them to conditioning and 'harshing' and, in the hands of certain security units and at certain locations (Abu Ghraib, Guantanamo Bay, Bagram Airbase, so-called 'black sites' around the world), direct physical and mental harm. Most of the Iraqis who suffered such treatment were admitted to be either innocent or ignorant, usually both. Thomas Ricks, an American journalist, reported that as many as ninety per cent of detainees in the first eighteen months of Iraq's occupation fell into these categories.* The lack of purpose and focus for intelligence gathering meant that thousands of Iraqis were rounded up, processed, and pushed out the other end with little or no useful information forthcoming. For some it was reminiscent of those desperate measures taken in Northern Ireland in the early 1970s: similar threat, similar solution, similar failure.

This was the stitch that bound Baha Mousa and the TQers and Anzio Company and the detainees to the war on terror declared by George W. Bush after September 2001. Did anyone in the chain appreciate the connection? Were the personnel of 1QLR or 19 Mechanised Brigade or the south-eastern Iraq division aware of their link to these strategic matters? Some maybe, those working in intelligence. But the troops on the ground, the ones who guarded

---

* Thomas Ricks' book *Fiasco: The American Military Adventure in Iraq* was published by Penguin in 2006.

the processed detainees, those who questioned them? It seems unlikely.
Perhaps that explains why it all seemed so aimless. No one really knew what
'intelligence' they were supposed to be seeking or why it was important.

32

ONCE THE STATEMENTS OF SSgt Davies and Sergeant Smulski had been taken, WO Spence returned to Shaibah HQ to see his commanding officer, Captain Nugent. The case was developing into a festival of documents with the compiled statements and orders and protocols and notes already numbering thousands of pages. There was still an unenviable amount of work to do, but the shape of the inquiry was attaining some coherence that had seemed unlikely a week before.

Spence handed over the file to his superior so that she could appreciate the state of play. When they met to discuss the paperwork on 30 September there was consensus about the future trajectory of the case. They identified four parallel tracks for the investigation to follow.

The first was obvious. Those individuals responsible for the violent death of Baha Mousa should be pursued: arrest followed by prosecution. Of course, the Army Prosecuting Authority would determine the direction and character of this path. They would have to be informed once the investigation was more complete. There was still quite a way to go before that happened. For now, though, they had a real suspect in Corporal Donald Payne. He matched the description given by SAC Hughes and both Riley and Betteridge as a main culprit in the violent treatment of all the detainees. He was in charge of the detention facility and he had already admitted to SSgt Sherrie Cooper that he was attempting to restrain Baha Mousa when the prisoner died. They also had Private Aaron Cooper's account which described Payne's brutal attack immediately prior to Baha Mousa's collapse. The medical and circumstantial evidence confirmed that Payne had to be the prime suspect. It was time for him to be taken into custody, cautioned and interviewed.

The second track was wider. All the witness statements of the detainees said that they had been subjected to numerous attacks and cruelties by a succession of soldiers and not just one man. These potential suspects could be divided into roughly four subsets: members of Anzio Company who might have assaulted the detainees both during the arrest on 14 September and during their duties as guards at the detention facility; the tactical questioners, SSgt Davies and Sergeant Smulski, who between them interrogated the detainees and directed some of their treatment; members of the provost staff who along with Corporal Payne were supposed to look after all prisoners on the base; and a host of other soldiers who'd visited the detention facility and randomly participated in the violence. Crimes may have been committed by members of each of these groups.

The third track was a little more complicated. So far the assumption was that something had gone deeply wrong over at Battle Group Main. It was highly unlikely that only a few foot soldiers and a number of guards were solely responsible. There was at least a suspicion that the actual abuse if not ordered was passively sanctioned. But by whom and to what degree? What orders had been given and what orders had been ignored? There would have to be considerable ferreting by the SIB before the scope of responsibilities could be drawn.

And finally there was a fourth track, perhaps even more challenging. The issue of the detainees' welfare had to be considered; could anyone say this was sufficient? Or did it amount to neglect? Was there a case to answer here? Who would be responsible? Those in the medical unit that served Battle Group Main, of course, but what about the general officer command? Could they be answerable for failing to oversee a proper system of care for the detainees? If so, what guidance about treatment of prisoners had they received from higher up the command chain? Davies' and Smulski's statements pointed a curling finger back to Britain, to the Defence Intelligence Centre at Chicksands. Had the whole scheme of training for interrogation been infected? Where else had the idea of 'harshing', 'conditioning', 'shock of capture' come from? Who was going to say these techniques were legal? Who would defend them? No one in the SIB could predict where this path could take them.

The SIB resolved to follow each strand as best they could. For now, though, resources were limited and they needed to focus.

Then Daoud Mousa turned up at the gates at Shaibah, where the SIB was housed, to exert more pressure. WO Spence went down to see him. He had to tell him that the investigation was likely to take months not days. There was no point in chasing them yet. Daoud Mousa asked to be a witness, but his offer was refused. After all, he hadn't actually seen his son being assaulted. Whether Daoud Mousa repeated what he'd seen at the time of his son's arrest at the hotel wasn't clear. Maybe it would have made a difference.

With Mousa out of the way, the investigation could continue. The death in custody remained the most obvious and immediate concern and they had now accumulated enough evidence to pick up Corporal Payne.

33

WHAT KIND OF MAN WAS Corporal Donald Payne? Few accounts by those who served with him failed to mention his physicality and the aura of violence he projected. Some said they were scared of him. He was regarded as a soldier to avoid. A 'bully' was one description. LCpl Adrian Redfearn felt intimidated by him. Private Gareth Aspinall spoke of his 'reputation'. Private Aaron Cooper said he was known as a boxer, one of the hardest men in the battalion. Whether justified or not, stories of a brutal and brutalising Donald Payne were commonplace from the moment the investigation into Baha Mousa's death began. But these were only stories. What else was known about him?

Payne wasn't one of those soldiers lacking experience. Unlike many of the men in A Company he knew what it was like to be in a hostile environment and faced by an unseen and dangerous enemy. Now in his mid-thirties, his career in the army had begun in 1988. He started off as a drummer boy and was proud to have played before royalty before moving to the infantry. He joined 1QLR in mid-1989 when he first came into contact with Jorge Mendonça, who was then the regimental adjutant. By the time they were in Iraq, the two of them had become quite familiar. They used to take early-morning runs together around Battle Group Main, so Payne remembered, up at 5.30 am for an invigorating jog about the compound.

But it was before Iraq that Payne had acquired his battle experience. His service spanned the latter and bloody end of the Northern Ireland Troubles. He completed several tours in the province and was in Omagh the day of the bombing which killed twenty-nine and injured a further 220. His wife and he were out shopping in the town and they missed the explosion by minutes. Hearing the blast, he went back to help and for most of the day attended the temporary mortuary. It would make you hard, something like that.

By 2001, Corporal Payne had joined the Regimental Police, ensuring the good behaviour of members of the regiment, guarding those locked up for breach of army rules. They weren't trained to handle civilians, though, Payne would say. That was a role only thrust upon him when he landed in Basra in July 2003. He learned the basics in Iraq, he said, from a tactical questioner soon after handover from the Black Watch who had occupied Battle Group Main prior to 1QLR. All that he did in the detention facility, the conditioning, the hooding, the stressing, he would say, had been in place then and he was expected, indeed ordered, to do the same.

On 29 September the SIB made arrangements for Payne's arrest. Sergeant Andrew Gordon was assigned the duty.

The speed with which Corporal Payne was arrested in Battle Group Main (at 9.27 am, 1 October 2003), his accommodation searched (by 11.15 am), taken to Basra airport and escorted on to an army transport plane and flown out to Britain (at 3 am, 2 October), conveyed on landing to Camp Bulford in Wiltshire (arriving at 9.30 pm on the 2nd), cautioned, provided with a legal representative, and seated, awaiting interview on his involvement in the death of Baha Mousa (by 11.44 am, 3 October), must have been dizzying. From the heat of Iraq to the dank cold of an English autumn. Little wonder that Sergeant Gordon, who had accompanied him throughout, should begin the questioning by asking whether Corporal Payne had had plenty of sleep and refreshments. A disoriented and jet-lagged interviewee might not provide the most accurate and considered information. But Payne had no complaints.

Gordon asked the corporal to fill him in on some background first. Payne told him about his appointment as provost corporal when he arrived in Basra during June 2003. He said he was made responsible for looking after Iraqis captured during 1QLR operations. These men would be held in the holding area, the three-roomed building they called the detention facility. Major Michael Peebles was the officer he reported to and Payne insisted he was never in charge of the guards unit. This was the responsibility of the operations officer, Captain Seeds. It was Seeds who would assign guards from the companies in Battle Group Main, usually members of the platoon who had made the arrests, dividing them into shifts, or 'stags', groups of two or more soldiers, to look after any detained Iraqis.

Corporal Payne said he knew nothing about Operation Salerno (when the detainees had been arrested), he hadn't been briefed about possible prisoners arriving and was only made aware that morning of 14 September. He recalled the prisoners had eventually appeared bound with plasticuffs. He had searched them and taken down their details and then had them spread between the two main rooms of the detention facility. The detainees were made to stand, but as time went on they would get tired and the guards would let them kneel. They would have their arms outstretched, he said, but not at shoulder height, more like waist level so their hands could be seen. Payne said the detainees had sandbag hoods over their heads. This was normal. The detainees would be given water and if they became tired they would be allowed to sit down cross-legged. Once every couple of hours, he said, each detainee would be walked around for some exercise. It all sounded quite gentle if Payne was to be believed.

Payne also said that the medical officer came and saw the detainees about three or four times to check on their condition. As far as he knew, there were no medical issues, no complaints from the detainees and no violence done to them by him or anyone else. The only problem was with the sandbags. He said that normally once a prisoner had been interrogated they would remove the hoods and let them relax a bit. After the first round of questioning, during the night of the 14th and early morning of the 15th, the tactical questioner (he didn't know the man's name) ordered the sandbags to be left on and the detainees to be kept awake. Payne said the TQer had told the guards to bang iron bars on the floor.

Then Sergeant Gordon asked about Baha Mousa.

'He was difficult,' Payne said. After he'd been subjected to TQing, he started getting out of his cuffs. Payne didn't know how, but he managed it several times. Payne said the TQer told him to cuff Mousa's thumbs, and when that didn't work, to cuff his hands to his ankles. That was hopeless too. So the TQer told him to put Mousa in the toilet area, the middle room of the detention facility. Eventually Mousa quietened down and Payne thought the problem was over.

Much later, at about 9 pm on the night of the 15th, Payne said he became worried that a relief for the guards on duty at the detention facility hadn't turned up. He'd checked the ops room and discovered the relief stag would be coming back to Battle Group Main shortly. When a Saxon truck containing the replacement guards finally arrived, he made his way over to the detention block, entering by the left-hand door. As he passed the middle toilet room he saw Mousa standing up with his cuffs off again and his sandbag removed from his head. He thought he was trying to escape. Payne said he pushed him back into the toilet area and got the man down on the floor. Payne said he had hold of the man's arm, and then Private Cooper and LCpl Redfearn had appeared. Payne said Cooper got hold of the man's other arm, the man was struggling violently. Payne said he put his knee in his back, but the prisoner pulled his arm free from Cooper's grasp. Payne said he grabbed both arms, applied more pressure, but the prisoner was still struggling and . . . and he banged his head. Payne said Cooper checked for a pulse, but Payne decided to call for the MO anyway. That was it.

'How did Mousa bang his head?' Sergeant Gordon wanted to know.

He was thrashing about, Payne told him, but his back was turned and he didn't see. He heard the crack, though. It could have been against the floor, it could have been the wall. With Private Cooper's assistance he quickly sat Mousa up, but his head was lolling to one side and his body sort of flopped and he realised there was something wrong. That's when they tried his pulse.

Sergeant Gordon asked whether Payne had kicked or punched Mousa or any of the other detainees.

'No,' said Payne.

Had he grabbed him around the neck, pulled his clothing about his throat?

'No.'

He couldn't tell Gordon how the detainees had received the bruising and injuries found later either. Payne did manage to say that he'd complained to the TQer about the orders to keep the bags on and not let the detainees sleep after they had been interrogated. And he said Major Peebles had been present when he'd said this. Payne thought the orders were wrong, but he had to follow them, didn't he?

Whether wittingly or not Corporal Payne had now implicated an officer. He may still have been in the frame for the death, but by his account the whole system had been imposed on him. Now there wasn't only the tactical questioners who were involved. The internment officer, Major Peebles, was too. So how far up the chain of command did this go?

34

W HILST Sergeant Gordon was interviewing Donald Payne, WO Spence was trying to keep a grip on the investigation. On 3 October the SIB team received a call from the brigade adjutant Captain Grogan, who wanted to know what was happening with the case. *The Times* had apparently been in touch, presumably with the Ministry of Defence, and were asking questions. Captain Gale Nugent was able to say there had been three arrests. It was enough to show the SIB were taking the matter seriously and acting with commendable swiftness. The story broke on 4 October, although it didn't attract much attention initially. Details were too sketchy to inspire any significant press outcry. The piece relied mostly on Daoud Mousa's demands for a proper inquiry. He had been in touch with *Al Jazeera*, determined not to let the killing of his son evaporate into the dust and sand of Basra. The story had been picked up by western news agencies, although it remained a minor item. Spence and the SIB didn't see the publicity as a problem, however. They believed they were already making the progress called for by Daoud Mousa.

Even so, it was plain to Spence that this was no time for relaxation. Each strand of the inquiry demanded action. Besides, information was gradually working its way to the team, crab-like in some cases. Major Quegan, the Territorial Army officer who had helped Captain Gareth Seeds attend to some of the detainees soon after Baha Mousa had died, had had a quiet word with SSgt Sherrie Cooper during one of her visits to Battle Group Main. He'd told her that he lived opposite the detention facility in the accommodation block and had been inside the building the night of Mousa's death. But he hadn't wanted to be contacted through his unit and hadn't given her a telephone number. He'd said he would contact her later to arrange a time when he would be at Brigade so they could meet without his fellow officers suspecting. Something was worrying him enough to go behind his normal chain of command. When eventually he found an excuse to visit the SIB, without letting his commanding officer know, he told the SIB that he'd always been worried about the way prisoners were being treated. He had seen men hooded, kneeling in the blistering sun whilst guards stood over them. He hadn't enquired what this was all about, but he had become increasingly upset. It didn't look right to him. He may have been a lawyer back home but he didn't need any legal training to believe this kind of treatment was wrong.

Although Major Quegan's account confirmed the nature of prisoner handling at 1QLR, it didn't really tell the SIB anything new. Nevertheless, it was another indication that the system in operation wasn't limited to a couple of provost staff at Battle Group Main. Ill-treatment appeared to have been open and visible to the whole camp.

What troubled WO Spence more immediately was the need for clear identification of suspects. Quegan couldn't help with that. So far they had a compliant witness in SAC Scott Hughes, the nervous teenager, but he was back in Britain on extended leave. They needed him to confirm who it was he had seen abusing the detainees.

Spence phoned around. He was told Hughes wasn't going to be brought back to serve in Basra. His time in Iraq was over. This provoked a small whirlwind of calls. Spence rang WO Cresswell, who was Hughes' boss, but only reached an answerphone. He then called Wing Commander Hunter who was in charge of welfare for Hughes' unit. Spence impressed upon him the necessity of getting Hughes back to Iraq. Hunter said he would contact his medical people and get back to him. WO Cresswell also called back and asked for a fax stating why Hughes had to be returned.

By 7 pm on Friday 3 October the formalities were completed and Hughes was scheduled to make the flight to Basra the following Monday. In the meantime, Spence set about preparing for a surreptitious identification procedure. As there were so many potential suspects, people whom Hughes had seen wander into the detention facility and assault one or more of the detainees during his hour-long visit there, it would be awkward to arrange any sort of identification parade. They could hardly line up the whole of 1QLR behind a one-sided mirror. Nor could they parade everyone and pass Hughes along them. It was vital that he was protected. Once it became known that he was a 'snitch' his life in the army would become difficult if not under threat. They were obliged to keep his identity secret, at least for the time being.

SAC Hughes was picked up from Basra International Airport the day after he had flown in on 6 October. WO Spence arrived with SSgt Sherrie

Cooper late morning on the 7th in a Land Rover with the rear-seat windows made of mirrored glass. Hughes was told to get in the back. They were going to Battle Group Main, it was explained, where they would park in the same place Hughes had parked when visiting with the *GMTV* crew three weeks before. They would wait for soldiers to come and go and he would identify any of those he'd seen in the detention facility. No one could look through the windows, Spence assured him. All they had to do was stay quiet and wait.

When they got to Battle Group Main and parked up, SSgt Cooper got out of the Land Rover and walked away from the vehicle. Spence and Hughes sat in the back. Within ten minutes a soldier, a 'big guy' wearing a purple T-shirt, appeared and passed behind the Land Rover on his way to the accommodation block. He was only feet away from Hughes.

'That's him,' Hughes said.

'Who?'

It was the tall and muscular soldier, a bull of a man, and the one who had made such a swift but violent visit to the detention facility whilst Hughes was there.

'Sure?'

'Sure.'

The 'big guy' disappeared into the building and then came out again.

'Is that him?'

'Yes.'

Spence rang Sherrie Cooper on a mobile phone and she emerged from elsewhere on the compound, approached the soldier and stopped him. He was arrested and led away. His name was Private Craig Slicker.

35

PRIVATE SLICKER'S IDENTIFICATION WAS IMPORTANT not simply because he'd been seen assaulting one of the detainees. He was important because he *wasn't* a member of Anzio Company. He worked at Battle Group Main in the Company quartermaster sergeant's stores, receiving and issuing equipment to the troops of 1QLR. He had nothing to do with operations in the field, hadn't taken part in Operation Salerno, provided no food or water for the prisoners and had nothing to do with guarding them. There was absolutely no reason for him to be anywhere near the prisoners or the detention block, other than perhaps to use the Portaloos outside. The fact that he had wandered in unchallenged and had lent a stiff boot to one of the detainees' kidneys suggested that the SIB's investigative net had to be spread wider than the soldiers of A Company.

Of course, they already suspected this. But here was the first proof that other soldiers on the base had been involved in the ill-treatment. It begged several questions: was Slicker's intervention an aberration? Or was it part of a behavioural pattern, where soldiers of 1QLR would take it upon themselves to drop into the detention block, gawp at the prisoners and perhaps even inflict some injury on them? And if it was widespread, how on earth could the officers of 1QLR have been unaware of what was going on? Could they have been that blind? If they were, what did that say about the nature of command? And was this incident unique? Had other detainees been treated the same? If not, what was so different about these men arrested at the Hotel al-Haitham?

SAC Hughes couldn't answer these questions. He was little more than an adolescent. But there was a burden upon him which was significantly greater than his years, and his work was not finished. WO Spence needed to identify the others Hughes had seen in the detention facility. Anzio Company was out on patrol in Basra and they would have to follow them on to the streets if they were to give Hughes the opportunity of seeing those faces again. The blacked-out Land Rover drove to the location where A Company were operating. An SIB officer had been sent on ahead, ready to move in when he received a message from Spence.

From the safety of the Land Rover's back seat, Hughes looked carefully at the patrolling soldiers. He saw someone he recognised despite the body armour and bulky helmet. It was one of the men from the detention facility. He remembered him playing a game with a prisoner. The soldier would click his fingers and the detainee would have to repeat a vulgar phrase.

When the Land Rover flashed its lights, SSgt Daren Jay of the SIB approached the soldier walking on his own, a little apart from his colleagues. Jay stopped him and checked his ID. It was Private Peter Bentham, a name that had barely registered with the investigators. He was just one of the guards at the detention facility, another member of Anzio Company.

'I'm arresting you', said SSgt Jay with painstaking formality, 'under section 74 of the Army Act 1955 on suspicion of being involved in the murder of Baha Dawood Salem and ill-treatment and torture of other internees at 1QLR Battle Group Main between Sunday 14 September 2003 and Monday 15 September 2003. You do not have to say anything but it may harm your defence if you do not mention when questioned something which you later rely on in court. Anything you do say may be given in evidence. Do you understand?'

'Yes,' was all Private Bentham said. But it must have felt strange, receiving the strict police formalities on a dusty street in downtown Basra, with the sun starting to go down, the heat still stultifying, a rifle in one's hands, wearing constricting body armour and helmet, being handed a booklet on the 'Rights of a Soldier' and asked to consent to one's accommodation being searched to collect clothing and boots worn at the time of the alleged offences. It must have been strange, like Corporal Payne before him, to be then whisked away from that street in Iraq back to Britain to an army barracks in the bleakness of a British country base suspected of some link with murder and torture.

Private Peter Bentham was a slightly unusual member of A Company. Originally an army chef, he left the forces in 1993 but then joined the Territorial Army in late 2002. He would say later that he then experienced infantry training for the first time. He passed the course and was accepted into the Lancastrian and Cumbrian Volunteers in the following March, the month of the Iraq invasion. He was called up in May and posted to Iraq with 1QLR in June. There was some induction in the few weeks leading up to his deployment, but it's difficult to see how he could have been presented as a member of an elite and experienced fighting force in one of the world's most dangerous and sensitive locations. Prepared or not, Private Bentham found himself on the streets of Basra along with the hardened soldiers of Anzio Company in September 2003.

Bentham wasn't a complete fish out of water, though. He was a friend of one of the other soldiers in the Company, Private MacKenzie whom he knew from back home, and seemed comfortable within the Anzio multiple. He was a willing team member. There was no 'civilian liaison' work for him. Patrol, guard duty, pursuit of suspects, these were very much part of his life in Iraq and fitted the commanding officer's ideal of a crack fighting unit. Perhaps it was fun for him too.

SAC Hughes picked Bentham out as one of the men he'd chatted to at the detention centre during the *GMTV* visit to Battle Group Main. Bentham was the one who'd told Hughes that the prisoners had pissed and shat themselves for good reason; they had been shouted at incessantly. Hughes' account suggested that Bentham was Male number 4 who had appeared very at ease with Corporal Payne dealing ferociously with the detainees. Hughes had also described Bentham squeezing water into 'Grandad's' mouth so hard that he couldn't swallow properly and had kicked the prisoners' feet as well to keep them in their cross-legged positions. It wasn't on the severe scale of ill-treatment, but even so the SIB decided to get him, like Payne, out of the country.

Private Bentham was taken to the airport and transported home. During his interview he denied any wrongdoing and he was soon released. In the absence of some kind of confession, there was simply insufficient evidence to connect Bentham to anything serious. Hughes' account was too weak. He hadn't seen Bentham kick anyone hard or treat them particularly roughly, certainly not to the extent that he could be accused of contributing to Baha Mousa's death or the severe injuries suffered by some of the other detainees. Later, Hughes would amend his evidence and say that the kicks he witnessed were little more than taps to the toes of the prisoners, taps designed to remind them to keep in position. Once that physical violence was removed from the picture there was little other than boorish behaviour which could be proven against Bentham. Hardly a war crime, the SIB must have decided.

But the arrest of Private Bentham wasn't a complete failure. It provoked some unintended consequences. Back in Basra, the Anzio Company members began to feel vulnerable. They knew about Corporal Payne's arrest and his sudden extraction from Battle Group Main. Now one of their own had been lifted from their midst. One minute he was on the streets with them, the next, he was under arrest, they thought, for murder. People began to talk.

36

7.30 PM, 9 OCTOBER 2003. Just as privates Slicker and Bentham were being processed by the SIB for transport to Britain and interview under caution, WO Spence received a phone call. Captain Mark Moutarde, the 1QLR adjutant, had rung to say that a number of men in Anzio Company wanted to speak to the investigators about the death of Baha Mousa.

It was no coincidence. The arrest of Pete Bentham had ignited a reaction in his comrades. But whether it was fear that they would be next or outrage that Bentham had been arrested wasn't yet clear. Had they been ordered to come forward or was it voluntary? Was it an attempt by 1QLR, and Anzio Company in particular, to take some control of a steadily deteriorating situation?

The whispers had already started before Bentham had been plucked from their sides. Corporal Payne had been spirited away as if merging into the early-morning desert miasma of heat shivers on tarmac. His presence was easily missed. That physicality and shockwave voice, the kind that make people wince and turn away, cross the road, anything to avoid contact, had seemed ever present at Battle Group Main: in the canteen, hanging about the accommodation block, in the quartermaster's stores, wandering the little world of the Battle Group Main compound. One couldn't fail but realise he had gone. It may have been a relief, but it was also a worry.

The rumours fanning out after his disappearance made people nervous. Conversations in quarters, a few words here and there; people must have known something was going on. The Anzio Company crew must have been drawing conclusions. And the officers couldn't have been immune from these developing feelings of dread either. Did they just sit back and wait? Or did they intervene in some way?

The men whose names Captain Moutarde put forward were Private Aaron Cooper, who had already given a statement when arrested the week before and was in the SIB's sights, Private Stuart MacKenzie, Bentham's friend in the Company, and Private Gareth Aspinall, who also worked closely with Bentham. They were anxious, and with good reason. All had taken part in the Hotel al-Haitham operation and all had been in the detention centre guarding the arrested Iraqis before Baha Mousa's death. If Bentham could be arrested for murder, as they supposed, and airlifted back home in a matter of moments, or so it seemed, they might be next. Was this their motivation for agreeing to talk?

Before any interviews could be arranged, Captain Moutarde rang again to say that he had another willing witness. Private Lee Graham, who at eighteen was one of the youngest in the battalion, had also come forward. The SIB picked up all four men and drove them to Shaibah Headquarters.

The first thing the investigators wanted to know was why the men had suddenly changed their story. SSgt Sherrie Cooper had seen them the day before, on 8 October, and all had denied witnessing any violence inflicted on the detainees by anyone. They had been close-lipped. What had changed? Only Gareth Aspinall offered an explanation. He'd been 'very scared', of one person in particular, he said. Now Corporal Payne was out of the way, he'd decided to tell the truth.

The justification didn't make much sense. Corporal Payne had been lifted on 1 October. Bentham had been arrested on the 7th. Aspinall and the other three had first been spoken to by SSgt Cooper on the 8th and had refused to say anything then about what had happened. Now, the next day, they had changed their minds. Simultaneously. The investigators were just pleased to break through the unit's silence so they didn't press for a more plausible explanation.

The story told by all four was pretty consistent. There was one common theme: it was all or mostly Corporal Payne's fault. He was the aggressor, the man who orchestrated the violence against the detainees. Everyone in Anzio Company was scared of him and couldn't bring themselves to say anything about the violence, couldn't and wouldn't do anything to stop it, to the point where they would even go along with the brutality, at least passively or with only mild involvement, so as not to provoke Payne or openly challenge him. They didn't seem embarrassed by their lack of courage.

The more the soldiers spoke the more they revealed about Payne's brutality and the involvement of a few others from outside their Company. Ripples of accusations against members of 1QLR appeared in their accounts. CSgt Robert Livesey's name came up. He was part of the intelligence cell in 1QLR and worked with the interrogators under the command of Major Peebles, the officer in charge of processing the detainees. On the first night after the arrest, during MacKenzie's 'stag' with Aaron Cooper, Sergeant Smith, who was Payne's immediate superior, had come into the facility and ordered them to let the detainees relax. Smith had told them they could take off the hoods and plasticuffs, give the prisoners water and let them sleep. About half an hour later, CSgt Livesey had come into the detention facility and seen the prisoners sleeping, without cuffs or hoods. He'd been angry and aggressive, MacKenzie said. Livesey ordered the guard to wake the detainees up, to take the water away, put the cuffs and hoods back on and get them standing. Then he'd gone again.

Private Lee Graham identified another visitor. In fact there had been 'a stream of unit personnel' wandering randomly into the detention block. Staff Sergeant Christopher Roberts, a senior member of H Company, was one of them. Graham described how Roberts had entered the building, aimed forceful kicks at three of the prisoners, who had cried out in pain, and then left. There had been no reason for the kicking, Graham said. He also remembered the doctor from the medical centre had come over to look at the old man and another prisoner who was complaining of shortness of breath.

Gareth Aspinall said that he'd seen a major, the interrogation officer (who must have been Major Peebles, although his name wasn't mentioned), come into the building and pick out a detainee for escorting back to questioning. It was this major who had also told them to segregate Baha Mousa and put him in the latrine. Baha was being such a nuisance by lifting his hood and somehow getting out of his cuffs, a story which supported Payne's account.

There were some admissions by the witnesses about their own behaviour, but they were couched in carefully guarded language. Graham confessed to 'slapping' the prisoners, but only as part of the shock culture preparing for interrogation. He said he, Bentham and Aspinall had all gently slapped the prisoners across the head occasionally. It was only intended to shock them into keeping in the stress positions. There was no intent to hurt them or cause them pain. Aspinall too said he slapped the detainees a couple of times around the face, but he was following his superiors' lead and thought it was just an 'insult' to Iraqis. He admitted that he'd struck a metal bar against the floor to keep the prisoners awake following the example of Payne and 'the major'. All four soldiers mentioned they had seen the 'choir' as well. It appeared to have been thought up and executed by Payne alone. It was his joke, they said. Aspinall found it funny to start with but towards the end of the second day he thought the prisoners had been through enough. Not that he did anything to stop it. He tried to spend as much time as possible outside with his book, he said.

It was Payne, then, who was labelled the chief and enduring culprit. He was the one who had instigated the 'choir' and who had enjoyed showing visitors how funny it was. He was the one who had kicked the prisoners with undiminished vigour. He was the one who had targeted 'Fatboy' (Baha Mousa) and 'Grandad' for particularly vicious treatment.

And there was another common thread. MacKenzie and Aspinall both claimed that Private Peter Bentham couldn't have had anything to do with the death. MacKenzie said that he'd returned to Battle Group Main from patrol at about 9.30 pm on 15 September and entered the detention facility to relieve privates Bentham, Aspinall and Graham on guard duty. MacKenzie saw the prisoners still in a bit of a state, slumped on the floor. He saw Corporal Payne in one of the main rooms and heard him shout his usual invective, 'Get your fucking heads up', punch a detainee about five times hard in the head, before seeing him walk into the middle room. MacKenzie said he went outside before returning and looking into the middle room, the former latrine. Private Bentham and LCpl Redfearn were shining torches into the room where Corporal Payne stood before a prisoner. MacKenzie saw Payne punch the prisoner in the stomach; he stumbled back against a wall. Payne grabbed the man by the top of his head and about the scruff of his neck and threw him to the floor. The prisoner hit the floor, face down. Payne stood over him with his arms pressed against either wall as though stabilising himself and began to kick the prisoner very hard. He used his right foot and kicked out six or seven times, MacKenzie said, striking the man about the head and shoulders. Payne was 'going berserk'. MacKenzie left the room only to return a few seconds later to see the prisoner lying motionless on the floor, Payne trying to pull him up and failing. The prisoner had gone limp, lifeless. Then everything went fucking mad, people shouting for a stretcher, for the medics.

A slightly different absolution was offered by Private Graham. He said that at 9 pm on the day of Baha Mousa's death, he was relieved of guard duty at the facility and jumped into the rear of a Saxon vehicle waiting to be taken back to barracks. Private Bentham sat in the back with him, which if correct, would mean he wasn't in the detention facility at all when Baha Mousa died. It took him away from the scene altogether.

Aspinall was in the detention facility then and he didn't mention Bentham. All he said was that he'd been waiting in the Saxon sometime after 9 pm when he'd heard lots of shouting and calls for a medic. He'd jumped out of the vehicle and tried to look inside the building to see what was going on. But it was too dark. Pitch black. Then the stretcher had appeared and 'Fatboy' was carried away. The damning part of his statement, though, was directed at Payne. Aspinall said he'd been in a 'huddle' with his mates outside the detention block after Baha Mousa had been stretchered away when Payne had appeared. He'd said to them all, 'If anyone asks, we were trying to put plasticuffs on and he banged his head' or something along those lines.

Aspinall took this to mean that he and the others were to keep to the script if asked any questions, which, of course, they had done for some time. Until now.

37

A LTHOUGH PRIVATE SLICKER AND OTHERS had been identified as 'visiting' the detention facility and MacKenzie, Aspinall and the others refuted any allegations of ill-treatment by them against the detainees, the men of Anzio Company were still the primary suspects for the assaults committed during arrest *and* detention. The SIB, therefore, had little choice but to interview the rest of the Company, particularly those in Lieutenant Rodgers' multiple. They had to immerse themselves in the culture of the unit, to understand what had gone on and see who else may have been responsible for the violence. They couldn't rely yet on the account that Corporal Payne was instigator and chief protagonist, dragooning or bullying others to take part in the 'conditioning'. Nor could they reasonably believe that Payne had acted as an authority with no senior command approval, particularly as so many people had seen what was going on in the detention facility and done nothing to prevent it. The SIB had known much of this from the first days of the investigation. It was only now that its detectives moved into Battle Group Main to get some answers. They had to be quick. The whole of 1QLR was due to decamp to Britain by the end of October, having reached the end of their short but bloody tour of duty.

Over the following two weeks all the members of Anzio Company were interviewed.

The first to be seen was Lieutenant Rodgers. He was becoming a pivotal figure for the inquiry. Many of the A Company witnesses were saying that they were either ordered to treat the detainees as they did or followed the example of others. Most of the blame was being laid on Corporal Payne. Private Aspinall was suggesting the intelligence officer, Major Peebles, had some responsibility and MacKenzie mentioned CSgt Livesey as telling his 'stag' what to do. None mentioned their multiple commander,

Lieutenant Rodgers. Was it plausible that his men had been directed to undertake guard duty at the detention facility without him having *any* sense of what they did there or the conditions they were supposed to maintain?

Lieutenant Rodgers' position was plain from the moment his interview began: he claimed he had nothing to do with the detention facility. As far as he was concerned this was the preserve of the provost staff, Sergeant Smith and Corporal Payne in particular. He was a line officer, no more, in charge of a platoon out in the field, patrolling, rapid response, armed operations. The army had its firmly defined chains of command and it wasn't for him to interfere in other people's duties. That would go against the whole structure of discipline which governed troops on active service or, for that matter, back home in barracks. Other officers were supposed to shoulder the tasks of looking after prisoners, questioning them, moving them on to other installations. He and his men were often required to help out, to provide guards and transport for Iraqis taken prisoner during his operations. But the care and treatment of detainees had nothing to do with him. The job was difficult and intense enough as it was. Naturally, his men had to pull guard duty from time to time. That was to be expected. He handed over his men to the provost staff to tell them what to do. It was like a loan. He would facilitate that but little else as far as he was concerned. His knowledge of tactical questioning or conditioning or any other of those 'intelligence' related issues was extremely limited, he told the SIB interviewer.

What about Operation Salerno? Was that different?

Well, yes, Rodgers said. He'd visited the detention facility when those prisoners lifted from the Hotel al-Haitham were held there, but only to oversee the handover between stags of his men. During one of these visits he'd noticed a detainee having 'breathing difficulties' and he spoke to him through Private Hunt, who apparently knew some Arabic. The prisoner said he wanted to see his father who was in the next room of the building, he was worried about him. Lieutenant Rodgers arranged for the old man to be brought to the entrance of the room so that the son could see he was OK. Apart from that Rodgers saw nothing to concern him. The place stank of body odour and it was very hot. Hardly surprising that. But he didn't see anything unduly bad about the place, certainly no violence. He knew the prisoners were to be kept awake prior to questioning and that they were being held in stress positions and had to wear hoods most of the time. There wasn't anything wrong with that, was there? It was all part of the softening-up prior to questioning. And it wasn't his affair in any case.

Did he know anything about the death of Baha Mousa?

Rodgers said he'd been at Battle Group Main that night, sitting with Major Peebles, the intelligence officer, for about thirty minutes receiving a briefing on another operation, when Corporal Payne had suddenly knocked on the door. He was visibly breathless and sweaty and told them that there was a problem with one of the prisoners. Major Peebles asked to be kept informed and Payne left, only to return within a few minutes to report that the prisoner was dead. Rodgers hurried to the ops room to make sure that a message about the death was sent through to Lt Col Mendonça. Major Peebles put on his combat shirt and supposedly headed off to the detention facility.

Once he had ensured the CO was informed, Rodgers decided to see what had happened with the prisoners. His men were there, after all. He walked over to the detention facility and quickly cornered Private Aspinall, who told him one of the prisoners had banged his head on a wall. There was some 'heated' discussion, Rodgers said, as he tried to get to the bottom of what had gone on. He was particularly concerned about why the man who had died had been kept in the latrine room. He didn't reveal why that troubled him so much.

Lieutenant Rodgers said the CO then arrived to say there would be an SIB investigation. A Company's officer in command, Major Englefield, also turned up. It was chaos. No one appeared able to say how the man had died or who was involved or provide any detail of significance at all. They were in the dark metaphorically and literally (there were only torches and distant street lights to illuminate the building).

Still in a state of ignorance, Rodgers organised his multiple. There were detainees to look after and he assigned a few of his men to continue with the guard duty. The rest he allowed to wander back to their Bedford trucks parked about Battle Group Main, to sleep where they could. In the morning, Rodgers and his men transported the remaining detainees to Camp Bucca. Interestingly, no mention was made then of what happened there, the uproar when the injuries to the prisoners were noticed or the direct accusations made against him by the reservist intelligence officer, Lt Cdr Crabbe.

That was all Lieutenant Rodgers had to say. Except for one final point, almost as an aside, but quietly tantalising nonetheless. He recalled that a few days after the death and the ensuing upheaval, probably on 6 or 7 October, he and his multiple were called to assist their sister Company, Burma, to help quell a riot that had threatened to break out in town. It wasn't an unusual occurrence and indeed the Burma Company sergeant major, Darren Leigh, was spectacularly familiar with such hazards. In August 2003, he and his company had been confronted by a large, hostile crowd, which outnumbered his men by ten to one. The situation had been on the cusp of becoming deadly. There had been rifle shots from the crowd, stones thrown, a couple of grenades. The 300 or so Iraqis had become emboldened with each act of violence directed at the small contingent of British troops. But CSM Leigh had refused to withdraw. Ignoring wounds to his legs caused by one of the exploding grenades, he had led his multiple in a baton charge which unnerved the mob and made them scatter. Afterwards, word had travelled around 1QLR and Leigh was seen as something of a hero. His commanding officer had recommended him for a decoration.

The situation wasn't quite so dramatic when Lieutenant Rodgers and his multiple had turned up to support Burma Company. Between the two companies they had managed to quell any outrage and disperse the crowd without too much difficulty. Afterwards, Rodgers said he stood chatting with CSM Leigh about the disturbance when quite without warning Leigh interjected to say the men of A Company were 'trying to fit up Don Payne'. Rodgers hadn't known what he was talking about and denied the suggestion, he said. Leigh allegedly retorted with a warning: if they say anything, then 'they'll get their comeuppance'.

Rodgers claimed he'd walked away at that point telling the man to 'grow up'. He'd later told his commanding officer, Major Englefield, about the comments, but despite otherwise keeping the matter to himself he soon realised, he said, that Leigh's remarks had reached his soldiers. They had become scared, he said.

If that was really true, why then had Aspinall, MacKenzie, Cooper and Graham all come forward within a day of the supposed exchange between Rodgers and Leigh? They can't have been that scared. The story suggested that either the men in A Company *were* lying about Payne, thus perpetrating a gross injustice and perhaps protecting someone else, *or* Leigh was trying somehow to shield Payne. It was a conundrum worth investigating.

CSM Leigh was asked about the alleged conversation, and the threat harboured within it, but not until February 2004 when he was stationed with the rest of 1QLR in Cyprus. His account was entirely different. Unsurprisingly, he denied any warning regarding Corporal Payne. He'd only made a sick joke as far as he remembered. 'I wonder who your multiple is going to kill today,' he'd said to Lieutenant Rodgers standing next to his Land Rover. Everyone had laughed. It was 'light-hearted' and 'friendly', the kind of banter that passed as humour amongst soldiers.

Why would Lieutenant Rodgers twist this conversation to suggest something more sinister? What purpose did it serve? And if Leigh was the one who was lying, what was hidden behind his threat?

The opportunity to look deeper into the conversation was lost a couple of months later. One Thursday in late April 2004, CSM Leigh was told that he'd been awarded a Military Cross for his actions in the riot in Basra the previous August. On the Friday, the decoration was made public. On the Saturday, he was dead. He had suffered a massive brain haemorrhage. It was his thirty-seventh birthday.

38

COULD THE SUDDEN DECISION BY some of the men in Anzio Company to speak to the SIB have been a wilful attempt to pile the blame on Corporal Payne? If so, it was a naïve tactic. Only if one presumed that the SIB were interested solely in how Baha Mousa had died, and perhaps how the other detainees had received their injuries, would sacrificing Payne satisfy the investigation. Much later, the first three who had come forward, privates Aspinall, MacKenzie and Cooper, would claim that Lieutenant Rodgers had put them up to it, had said they needed to go to the authorities and give an account of how Corporal Payne was *the* man responsible. Whether true or not, and Rodgers has always denied the allegation, anyone believing that a scapegoat would satisfy the SIB must also have thought that there was little if anything wrong with the whole system of treatment of prisoners and the supervising command structure. But the case had already moved away from the rogue-soldier story. Lines of inquiry were focused on the responsibility of officers to guide and police their men competently. *This* was the worry that had wormed its way through the military and bureaucratic channels and was irritating the conscience of public officials.

For now, though, there was still a case to be made against anyone who may have assaulted Baha Mousa and the other prisoners. So, whilst Lieutenant Rodgers was giving his version of events, other members of his Company were speaking to SIB investigators. Statements were drafted for a number of men on that same day, 12 October. They were of mixed value.

Private Thomas Appleby was one of the first to be interviewed. He described how he'd first entered the detention facility on the night of the initial arrest. He said it was dark and the smell inside was foul. He could make out Corporal Payne and two men from the Hollender multiple of Anzio Company, privates Wayne Crowcroft and Darren Fallon. Appleby knew Payne from his time back in Catterick; he had 'a bad reputation as a psychopath' although he didn't say why. His statement then depicted the same litany of crude abuse and sporadic but sustained violence which had become familiar in all the accounts so far. Appleby said he thought the treatment was 'harsh and horrible', but Payne's aggressive demeanour didn't encourage any confrontation. Appleby didn't have the courage to say anything against the treatment then or later.

As he came and went over the next thirty-six hours, performing stags as pretty much everyone else from Anzio Company had to do, Private Appleby saw the 'choir' orchestrated by Payne and said it had been performed in front of Lieutenant Rodgers along with Aspinall, Bentham and a couple of others. An officer had appeared at some point to tell him to keep the detainees awake, particularly the eighteen-year-old boy. They had to put the boy in the latrine, the middle of the three rooms. The officer reportedly told Appleby and his buddy on the stag Private Garry Reader (it was early morning on 15 September by then) that this lad was about to break and the questioners needed them to make as much noise as possible, to keep grabbing him and making him stand up, to march him outside and then back again.

And there were other scraps of information. Appleby saw Baha Mousa being taken away on a stretcher and then Private Reader appeared outside the detention facility, vomiting into the roadway. Reader had told him that he'd been giving CPR, but the prisoner had died anyway. Then Corporal

Payne had wandered up to them, very calm and 'unfazed' and said to the small group of soldiers that 'if anything comes of this, he banged his own head against the wall' or something similar. It confirmed Aspinall's account. There was little else Appleby had to offer.

Corporal John Douglas was one of those Territorial Army members serving with the Company. He was in his forties and, unlike most of the others in the multiple, didn't come from Lancashire. He would eventually join the regular army, but in Basra he was a volunteer, driving one of the Saxon vehicles on operations. Even though he was never assigned to guard the prisoners at Battle Group Main he admitted entering the detention facility a number of times. Indeed, he seemed to have assumed a role for himself, helping out with maintaining the prisoners' stress positions, giving them 'an encouraging slap around their arms' or 'a nudge with my foot on their feet'. He would shout at them too, he said, something like 'Get your hands up'. This wasn't 'physical abuse' as far as he was concerned. But he did see others using force, people who would come in and punch and kick. And, significantly, he claimed he was present when the final struggle with Baha Mousa took place in the middle room of the detention building. Douglas said he was outside the little room and watched as Corporal Payne and LCpl Redfearn tried to restrain the prisoner. He said he saw Payne punch and slap Mousa several times. Then, for some reason, Douglas went to his Saxon to get a Maglite torch. He returned to the room and shone the piercing beam inside. It must have cast a freezing white light on Payne punching the prisoner who Douglas thought appeared to be 'thrown' across the room, hitting the wall and banging the left side of his head before collapsing on to the floor.

Douglas said he shone the torch beam on the prisoner's face, which looked swollen, but with no blood as far as he could make out. He told Payne 'He doesn't look good, he looks dead to me.' Private Reader moved in to try resuscitation. It wasn't long before some medics appeared to take the body away.

Private Damien Kenny, by contrast, could remember next to nothing. There was no doubting that he'd been in the detention facility at various times. He'd been assigned to the same stag group as Cooper, Appleby, Allibone and MacKenzie, but he still had no recollection of anything that took place inside. The day after the SIB saw him he was hit by shrapnel in the hand, the result of an accidental discharge of a weapon, and had to be evacuated for treatment. His memory of all events surrounding the operation in the hotel or the detention facility back at base would never return, or so he continued to affirm.

Private Garry Reader had a lot to say though. Throughout his statement to the SIB he used nicknames for the detainees: 'Grandad' for the old man, 'Bruise' for the one who later suffered kidney failure, 'Fat bastard', 'Young guy', 'Pisspants' and with tasteless after-knowledge 'Die' for the prisoner he knew was killed, Baha Mousa. It was reminiscent of the names given to prisoners by US guards at Abu Ghraib. There, it had turned detainees into 'cartoon characters, which kept them comfortably unreal'.* And, by implication, susceptible to abuse as 'things' rather than human beings.

Reader was also free with his nicknames for others in his multiple: Private Allibone was called 'Haribo', Hunt was called 'Gorgeous' (because he was so ugly), and more favourably, but less humorously, 'Aps', 'Mac', 'Coops', 'Redders'.

Reader saw various people in the facility on the evening of 14 September when he came to do his stag. Payne was there and privates Fallon and 'Crowy' Crowcroft from the Hollender multiple of A Company. Fallon told him the score and showed him the prisoners, some of whom were already clearly bruised and beaten, Reader said. He was adamant, though, that no one in his stag had hit the detainees: no kicks, no punches, nothing other than trying to keep them in their stress positions as they had been ordered to do by Payne at the beginning of the shift. Yes, he was there at the death of Baha Mousa, or 'Die' as he called him. He saw Cooper and Payne grab him and push him back into the middle room of the block, but he didn't follow. He ignored the screaming coming from the room and went to stand outside. He talked with Private Graham and told him what he'd just seen. Then the screaming stopped. He went back in and looked into the middle room where 'Die' was propped against a wall. Someone said 'He's in trouble.' The man wasn't breathing and Reader rushed in to try to revive him. He placed him on his back, checked for a pulse (there was one, but it was weak) checked his airway and started mouth-to-mouth resuscitation. Then the medic arrived, he said. No mention was made about throwing up afterwards.

* The nicknames were noted in Philip Gourevitch's and Errol Morris' investigative work, *Standard Operating Procedure: A War Story* (Picador, 2008, page 100).

Private Jonathan Hunt had just turned twenty-one. He was Lieutenant Rodgers' signalman, responsible for communicating any information from his commander by radio to HQ. During the 'soft knock' against the Hotel al-Haitham, he witnessed the hotel employees being arrested, but had nothing to do with their treatment there. The names Fallon and Crowcroft came up again as two of the men ordered to escort the detainees from the hotel back to Battle Group Main.

Hunt described how he, along with others in his multiple, privates Stirland, Bentham, Graham and Aspinall, were delegated to carry out guard duty in the morning of the second day of detention. Lieutenant Rodgers briefed them informally, telling them not to let the prisoners talk or sleep, to shout at them, but not to use physical violence and to make sure they had water. Apart from keeping the prisoners cuffed and hooded, Hunt confirmed neither he nor Stirland, his guard duty partner, saw any trouble or violence. Stress positions weren't maintained, he said. He did see Lieutenant Rodgers bring in a metal bar and drop it loudly on the floor, but that was about it. Several people visited the facility during his stag, popping in after using the Portaloos outside. None of these men, neither Corporal Payne nor anyone else, assaulted anyone, he said.

On the evening of Baha Mousa's death, Hunt arrived along with a number of his Company at about 9.30 pm. They had been out on patrol but were then on their way to relieve the guard. He had seen a melee of soldiers, perhaps twelve or more, congregated outside the middle room. He'd asked what was going on and was told that one of the prisoners had collapsed. Hunt pushed his way through, he said, because he'd received some first-aid training in the past. Private Reader was already trying resuscitation so all Hunt did was help the medics carry the collapsed detainee to the medical centre. After that, he returned to the facility, but was then assigned to act as 'top cover' for the trucks taking Baha Mousa's body to the mortuary at Shaibah. On his return from this macabre duty he'd slept for about three hours and then took over guarding the remaining prisoners for a while. In the morning he and the rest of the multiple still on site gave the prisoners a bit of a workout, getting them to do a few exercises and stretches, nothing strenuous, he said. He couldn't remember why this had been ordered.

This was as much as the SIB officers could extract on 12 October.

On the 13th more soldiers were interviewed. Corporal Dawson from the Hollender multiple had little to say. He hadn't performed any guard duty at Battle Group Main and hadn't entered the detention facility. The same applied to Lance Corporal Stephen Woods. His statement was short and uninformative. Private Andrew Altree was a raw recruit, only joining the army the previous September, and he also had nothing to tell the SIB. It seemed most of the Hollender multiple of Anzio Company could be excluded from the inquiries. They hadn't been involved in dealing with the detainees and hadn't even spent time at Battle Group Main during 14–16 September.

There were three exceptions: Corporal Stacey, Private Crowcroft and Private Fallon. They had all been mentioned with notable frequency as being the first to have guarded the detainees immediately after the morning arrest on the Sunday. Members of Lieutenant Rodgers' multiple had been kept back from these duties until later that day, arriving early in the evening, coming as a pack to the detention facility and only then taking it in turns to do their stags. Some of these men had seen the detainees already in a state when they arrived. It suggested that Stacey, Crowcroft and Fallon may have had something to do with their injuries.

But the SIB couldn't interview Corporal Stacey. He'd returned to Britain to get married. Gone were the days when such personal commitments would be jettisoned when on active service. The army was a very sympathetic employer now.

Crowcroft and Fallon, on the other hand, were available. They were called in and interviewed separately. By the end of the day the SIB had two statements. They were remarkably consistent.

Both soldiers said they had been lumbered with looking after the prisoners on that first day and hadn't been best pleased about it, left as they were with the responsibility for most of the morning and afternoon. Hot and tired after the early-morning antics of Operation Salerno, they felt aggrieved at having to guard the prisoners from 11 am until about 6 pm. It was a dangerously long shift. And they didn't appear particularly well prepared for such a role. Both were barely out of school.

Indeed, Wayne Crowcroft would have been called a child soldier in many armies of the world. He joined up at the age of sixteen and was posted to Iraq with 1QLR only two years later, by then a trained heavy machine gunner. Darren Fallon wasn't much older, no more than twenty. It was a very delicate age to be thrown into the chaos of Basra in the aftermath of war and the beginnings of the violent local resistance to occupation, or 'insurgency' as it was becoming known. But most of his company were the same. The non-commissioned officers may have been older, but the bulk of them were all very young, eighteen, nineteen, twenty. The mature rank and file tended to be Territorial Army recruits who came from other occupations. It made for a strange mixture of trained but immature and mature but inexperienced soldiers. How prepared could they have been for the streets in Iraq, where shots from darkened alleys and suddenly violent crowds were to be expected, and for the guarding of a group of suspected terrorists, as the Iraqis captured at the Hotel al-Haitham were described?

Crowcroft and Fallon each spoke of the lengthy guard duty they'd had to endure, the initial briefing by Corporal Payne, who told them to keep the detainees in stress positions but to give the prisoners water when needed, and the use of handcuffs and hoods despite the heat. Those were their orders. They denied any violence was inflicted by them or Payne or anyone else. The only time there was physical contact was when they 'assisted' the detainees to keep in their stressed poses. There was one incident, though, which both recalled in very similar words. One of the prisoners had suddenly leapt at Private Fallon, just as he'd been giving him water, and pushed him to the floor. Crowcroft had grabbed the detainee by the neck, pulled him off his colleague, and subdued him. He hadn't used unreasonable force in the circumstances, Crowcroft said. Fallon agreed. And when their shift had finally ended, the detainees had been handed over uninjured. They were adamant about that.

39

TIME WAS RUNNING OUT FOR the SIB team. The slew of statements collected made it look as though they had made good progress, but in truth they were nowhere near finalising their inquiries. It was now the middle of October and 1QLR were preparing to leave Iraq. The investigation would have to follow them home, a messy prospect. Troops would be scattered, documents mislaid, arranging identification parades would be complicated. Keeping track of witnesses, particularly those in the Territorial Army discharged on their return, would be difficult. It would be an administrative puzzle which could only delay the investigation and might even undermine any prosecution that followed. But there was little they could do about that.

Most officers now suspected that the matter would take months to resolve. There were so many interweaving stories, from almost random acts of low-level violence (a punch here, a kick there, which although objectionable hardly represented the basis for a major war-crimes trial), to those of more sustained cruelty. But if men other than Donald Payne were to be prosecuted, much more needed to be done. Formal ID parades, corroborating evidence, hopefully confessions and direct witness testimony – all would need to be gathered for delivery to the army lawyers. Captain Gale Nugent believed that they wouldn't be finished until at least February 2004, and that was an overly optimistic estimate.

WO Spence, who had day-to-day control of the inquiry, was also due to return home on 26 October. This complicated the investigation even further. It meant he only had a few days from finishing interviewing the men of A Company until he would have to pack up the case files. Before he left he wanted, at least, to establish how responsibility for the detainees was assumed by the chain of command. This was the second strand of the SIB's investigation. Who in the senior command knew or should have known what was going on?

There were a number of officers who hadn't yet been interviewed by the SIB, and Spence made these his priority.

First on the list was Major Robert Englefield. He was the commander of A Company and in overall charge of Lieutenant Rodgers, CSgt Hollender, and their respective multiples. But Englefield was strangely absent from any of the accounts collected so far. Spence went to see him on 20 October.

Major Englefield admitted he had been centrally involved in the planning and conduct of Operation Salerno, but he couldn't say whether anyone was ill-treated during the arrests in the hotel. He'd seen seven hotel employees lying prone on the floor of the al-Haitham after his soldiers had discovered a small cache of arms, and he explained that they had arrested these men because they couldn't identify which one was responsible for the guns. Better to take them *all* in for questioning than miss a possible terrorist. Englefield said he ordered the men to be plasticuffed, but not hooded. He didn't think it would 'look good in the media' for them to be seen herding a bunch of hooded men on to the Bedford truck parked outside the hotel.

Once the detainees were taken away he never saw them again. The next he heard was that one of the prisoners had died. He'd gone over to Battle Group Main then, but only to say to the men in his Company that 'if they required advice or assistance' they could come and see him. What he meant by this was obscure. What advice did he think they would need?

He then said he'd tried to find out what had happened. He spoke briefly with Lieutenant Rodgers and a number of the more senior men in the Rodgers multiple: LCpl Redfearn and privates Cooper and Graham. They told him that the prisoner had broken free and shoulder-charged someone or rugby-tackled them and when the man was returned to custody it was noticed he was not breathing. Major Englefield didn't enquire any further.

None of this made much sense now. The rugby tackle sounded very familiar, but that had been in the story recounted by privates Crowcroft and Fallon about an incident on the first day of detention, at least thirty hours *before* Baha Mousa's death. Now it was conjoined with the moment when Baha had died. But surely all the men Major Englefield said he spoke to knew this? Cooper, Redfearn and Graham were all there at Mousa's death.

There was little else Major Englefield had to offer. He did make one simple admission: he accepted that he hadn't questioned anyone further to determine who was responsible for what had happened. His ignorance appeared deep and his evidence pathetically thin.

Next on Spence's list was the commanding officer himself. Although Lt Col Jorge Mendonça had been kept well informed about the investigation and its various twists, he hadn't yet been interviewed. He may have been left alone whilst the more direct evidence of ill-treatment was collected, but it was strange that he hadn't been quizzed about his command responsibilities as well as his particular knowledge of the whole affair until over a month after the event. After all, the issue of culpability possibly stretching into the higher echelons of the officer corps had been identified by the SIB very early in their investigations. It was only now, 22 October, that an interview was finally arranged.

Even then, as with Major Englefield, the information extracted from Lt Col Mendonça was negligible. Perhaps the SIB officers were rushed. Perhaps they didn't know what to ask. Perhaps those interviewed were allowed to dictate their account without being probed. Perhaps deference was paid to those of more senior rank. Whatever the reason, as the investigation developed over succeeding months, indeed years, the inadequacy of these initial statements became apparent. The delays made a mockery of one rule of witness evidence collection: try to get a person's testimony committed to writing as close to the event as possible. It's a way of memorialising evidence.

Lt Col Mendonça began by talking about the battalion's guidelines for internment. They had been produced by Major Royce in July 2003. Royce had left theatre soon after and was replaced by Major Michael Peebles, but the guidelines remained standard operating procedure. Mendonça handed a copy to WO Spence.

He then spoke about Operation Salerno and the subsequent detention of the Iraqis arrested at the hotel. Of course, he knew a small number of hotel employees had been arrested. He had been there. And after they had all been transported back to camp he had kept himself up to date on the intelligence produced from tactical questioning by asking Major Peebles how the interrogation was getting on. Nothing much, was all he was told. He remembered that he briefly visited the detention facility on that first evening on Sunday 14 September. He said he wanted to ensure the prisoners had water. They had. He didn't stay long, he said, and left after a brief discussion with the guards. The detainees were seated and quiet, he recalled.

Then a strange, almost tangential episode was mentioned. Lt Col Mendonça said that early on the Monday morning, the 15th, he was having breakfast when he saw a soldier carry a stack of plates laden with food out of the canteen. They were piled with scrambled eggs, tomatoes, usual fare except no bacon, no sausages. Mendonça stopped the soldier and asked whether the food was for the guard. 'No' was the reply. It was for the prisoners. Mendonça couldn't remember the soldier's name.

It wasn't clear what the lieutenant colonel meant to convey with this little anecdote. Was it to show how well the detainees were being treated? That they were getting the same food, less the meat, as his men? Or was it an indication that they weren't being treated well at all, given that the reason for quizzing the soldier, he said, was because he was worried about hygiene? It was very odd. And he had nothing much else to tell WO Spence. He mentioned Baha Mousa's death but only to confirm how he had found out (whilst on patrol over the radio) and how he had ordered the SIB to be called in.

The statement produced was a pitifully short declaration of the commanding officer's knowledge. As the investigating team had already identified command responsibility and institutional neglect as important lines of inquiry, the statement indicated a flaccid attempt to mine any useful information. Where was the detail about his relationship with Major Peebles and his provost staff? Or the account of his actions once he'd heard about the death of one of the detainees? Maybe the lieutenant colonel had been busy and could only spare a short period of time to be interviewed. That wouldn't have been out of character for Mendonça's action-man style of command. But there was no note on file to say further inquiries would be necessary, that there was insufficient time to explore avenues of interest. Later, WO Spence would claim that he had been 'concerned to establish where responsibility lay within the chain of command for individuals within 1QLR custody' and to make sure he had the internment orders guiding the battalion. It was dubious whether he had succeeded in doing even that.

The guidelines for internment handed over by Lt Col Mendonça did, however, provide some idea about the system in operation. They made a distinction between those arrested who posed a threat to the British Army 'mission' in Iraq (these were called 'internees' and covered those intent on fighting Coalition troops) and anyone else. If someone wasn't identified as an internee then they had to be released or handed over to the Iraqi police if suspected of a normal crime. Otherwise, if classified as an internee, they had to be brought to Battle Group Main within two hours, handed over to the internment officer and taken to Camp Bucca within fourteen hours of arrest. Tactical questioning was not to be handled by the arresting Company, but they would be responsible for guarding those arrested. The obligations were clear, if limited. The 'fourteen-hour rule' was the one most obviously broken in the case of those arrested at the Hotel al-Haitham. It took more than forty-eight hours for them to be delivered to Camp Bucca, as all the records confirmed. Lt Col Mendonça's statement had little to say about this obvious breach.

Mendonça would be allowed to return to Britain at the end of October with the rest of the battalion. Many lines of inquiry remained outstanding. They would continue to shadow the officer for years.

40

*T*HE STATEMENT TAKEN FROM LT Col Mendonça confirmed a growing suspicion that the official procedure followed by his battalion was threadbare, inadequate as any kind of protocol which officers and men responsible for handling detainees could sensibly follow. It was wholly lacking *in recognition of those detailed protections which prisoners of war were supposed to enjoy.*

By now high command knew this very well. In trying to plug the obvious hole in the legal framework outlining how prisoners should be treated, senior members of the legal corps produced a letter emphasising what should now take place. They said that hooding 'is to stop'. It was categorical.

It wasn't just the killing of Baha Mousa which had prompted the change in orders. The letter sent to commanders in Iraq spoke of the 'hooding of prisoners of war, internees and detainees by UK military' having already 'attracted the attention of the international media and the International Committee of the Red Cross'. 'Adverse comments' about Britain's handling of prisoners had 'prompted an examination of long-standing practices by UK forces in prisoner handling'.

What did this all mean?

Far from being 'surprised' by the circumstances of the Baha Mousa case, it soon became clear that the army were more than familiar with such behaviour by its troops. It had been made known ever since the invasion. And not only were they aware, they were told in the clearest terms by the International Committee of the Red Cross on 1 April 2003, nearly six months before Baha Mousa's death. The ICRC had spoken to the political advisor of the commander of British armed forces in Iraq about 'methods of ill-treatment used by military intelligence personnel to interrogate persons deprived of their liberty'. The complaints related to the internment camp at Umm Qasr. 'Brutal' was the word the ICRC used to describe the systematic use of hoods and flexicuffs. Its on-site observer complained about prisoners being made to sit in the sun, about soldiers 'kicking' and using stress positions.

The political advisor reported what the ICRC had said to the Ministry of Defence and Air Chief Marshal Burridge. An ameliorating message was sent to the Secretary of Defence's office in April 2003 saying that immediate action had already been taken. The advisor would later recall in a memo that 'bagging' was ordered 'to stop forthwith as was harsh treatment'. The military chain of command was notified, the memo said, and 'closer supervision arranged'. Blindfolds or goggles were to be used instead of bags. Everything that needed to be done had been done, it said.

Such assurances to the ministry were to be expected. A condemnatory report by the ICRC such as this would be a political disaster if it came to light. Any decent political advisor would have known that. But fortunately for the British government and the army, the ICRC operates according to a *code of strict confidentiality. Its mission may be 'to protect the lives and dignity of victims of armed conflict' but its standard operating rule is to keep private whatever it finds. It believes that it can help detainees who are victims of ill-treatment best through negotiation, not through naming and shaming. By observing confidentiality it can gain access to prisons and internment camps and persuade state authorities to change abusive practices. In that way individuals can have their suffering relieved immediately. It doesn't matter that the world may never find out about cases of abuse. Helping victims is the priority. Of course, if a state doesn't cooperate then the ICRC may go public. Given its highly regarded status this often means what it says is taken as gospel. Knowing that a negative ICRC report can be politically damaging frequently forces states to respond positively to the complaints. That at least is the theory.*

In October 2003 the British military knew all about the ICRC concerns from the previous April. The legal experts knew. The political advisors knew. And the government knew. Yet, six months after measures had supposedly been undertaken to accord with international standards of treatment of detainees, here they were faced with an incident that repeated, almost abusive practice by abusive practice, the complaints previously highlighted by the ICRC. Had procedures ever been changed as promised? Had units new to theatre been properly briefed? Either way, the lack of oversight and control by the officer command was startling. Lt Col Mendonça's statement was yet another indication that the institutional failings seeped through every pore of the command structure.

But if it was true that neglect of duty with regard to ensuring the safety of detainees was systemic, how could the SIB identify individual officers responsible? If the command was tainted from headquarters in Britain down through Division, Brigade, Battalion and Company levels, where would the blame stop? They couldn't prosecute the whole army.

41

WO Spence already knew that the training of interrogators was an issue for the investigation. He'd already heard from SSgt Davies and Sergeant Smulski that they had encouraged sleep deprivation and hooding and stress positions because this was what they had been taught to do. If this was true, then practices condemned by the ICRC weren't the product of 'bad apples'. They were institutionalised, part of army teaching.

On 13 October 2003 Spence called Sergeant Gordon who was then still in Bulford, Wiltshire, after having recently finished interviewing Corporal Payne. Spence told Gordon he was worried about the information they had uncovered on the techniques of tactical questioning. Smulski and Davies had suggested that they had followed the training they had received at Chicksands. Sergeant Gordon was asked to make the journey to Bedfordshire to find out exactly what was taught there. Did they really train men to enforce conditioning? Were all those techniques of stress positions and 'harshing' and sleep deprivation the subject of classroom preparation? Gordon's report finally reached Spence on 5 November.

The location of one of the British Army's most sensitive departments isn't a secret. Chicksands has a working web page, advertising its attractions (it has a museum of intelligence-gathering which can be visited by appointment) and explaining its role in the modern British Army. The place is described as 'one of the most striking military bases in the UK', its focal point being a twelfth-century priory which has undergone many stately extensions over the years. A brief history applauds it as home now to the Intelligence Corps and its rich traditions of intensely dangerous operations in all conflicts where the British Army has been deployed during and since the Second World War. Chicksands was one of those centres for decoding secret German messages intercepted through the Enigma machine. After the war the US Air Force maintained the base as home to its intelligence operations. It was handed back in 1997. The Intelligence Corps now inhabits Chicksands as a training facility with particular emphasis on human intelligence. 'People', the website declares, 'are the Army's most valuable assets.' This includes not only army personnel, trained and ready to fight, but also, rather curiously, 'prisoners of war and civilians'. They are 'assets' too, being an 'invaluable source of intelligence'. Nothing is said about how this intelligence is to be extracted.

Sergeant Gordon's brief was to find out how Chicksands was training people to interrogate these civilian assets. He took a statement from the senior instructor of human intelligence skills, which included tactical questioning, on 14 October.

The officer provided somewhat ambiguous information. He told Sergeant Gordon that stress positions weren't taught on any course he organised and were actively discouraged. Plasticuffs were recommended, but hoods not. Blindfolds could be used. Maintaining the 'shock of capture' was not a specific aim of any system of prisoner handling as far as he was concerned. But he confirmed that various 'pressures' both self-induced and 'induced by the system and the conditions of being confined' were acceptable in encouraging a prisoner to talk.

A number of approaches were taught when it came to questioning: 'friendly, firm and logical and harsh'. The latter would only be used, he said, to make prisoners appreciate their situation. It wasn't intended to garner information, but rather appeared to be some kind of preparation for interview. The officer was adamant that no violence should be used at any stage. That wasn't part of the training.

There was little here to confirm Davies' and Smulski's stories about their training. Only one document suggested that they had ingested some training material: it was a directed method for moving prisoners from one location within a base to another. A paper described how the prisoner should extend his arms, linking hands together with thumbs uppermost, whereupon one guard would grip the upraised thumbs whilst the other would stand behind the prisoner and place his hands on the prisoner's shoulders. The prisoner would then be guided about the detention facility in this fashion. This was almost word for word Sergeant Smulski's account of procedure he followed in Basra prior to interrogation. And yet when the Chicksands officer checked the records he could find nothing to confirm that Sergeant Smulski had even attended one of his courses. SSgt Davies, yes, Smulski, no. And every tactical questioner was supposed to be trained.

Although the training officer left Sergeant Gordon with the clear impression that everything was done by the letter of the Geneva Conventions, what was the reality? What was said when the PowerPoint presentation had finished and soldiers were asking 'Yes, but what's it really like?' What did 'harsh' questioning mean? What did it look like? And how was it delivered? The training documents didn't explain. No doubt like any process, it was something that had to be seen, experienced, talked about, and practised. The two-week course that Davies had attended must have done more than sit trainees in a classroom and talk about 'harshing' as though it were possible to convey simply by using words. There must have been some workshop or role play, one of those methods to help translate the written guidelines. And if not, what kind of lessons could have been conveyed?

Much later, evidence emerged about what the training was like when the participants moved away from the blackboard. Some papers came to light describing the approach to prisoner 'handling' which interrogators were supposed to adopt towards detainees. One said 'get them naked' when taken to a secure holding location. Get them naked and 'keep them naked if they do not follow commands'. A more complete training manual said that a search and reception area should be provided which possessed 'an element of unspoken threat' and 'degree of privacy (discretion for arab modesty!) for full strip searches'. Perhaps it's possible to read too much into punctuation, but that exclamation mark alone resonated with contempt, a culture and attitude soaked in disdain.

When 'harshing' was mentioned, a lack of enthusiasm for its effectiveness was evident, but it wasn't banned. There was just caution that it didn't work. One manual warned: 'If you kill or injure a [prisoner] you cannot get information from them. You may be called to account for your actions as a war criminal.' May?

Some transcripts of a video also surfaced which demonstrated harsh techniques. Whether it was to show what shouldn't be undertaken is unclear. One extract shows an interrogator standing in front of a person playing the role of a detainee experiencing tactical questioning:

Interrogator (*shouting*): YOU ARE FUCKING STARTING TO PISS ME OFF, FUCKING NOBHEAD, FUCKING START ANSWERING QUESTIONS. STOOD THERE LIKE A FUCKING PIECE OF SHIT, THINKING YOU'RE THE FUCKING MAIN MAN. YOU'RE NOT THE FUCKING MAIN MAN, I ASSURE YOU, I AM THE MAN IN THIS FUCKING ROOM. IF YOU DON'T START HELPING, YOU'RE FUCKING SET. START ANSWERING THE QUESTIONS. STOP CHEWING YOUR FUCKING GUM, STOP THINKING YOU'RE FUCKING HARD CORE. STOP FOLDING YOUR ARMS, PRANCING AROUND. LOOK ME IN THE EYE AND START ANSWERING MY QUESTIONS. DO YOU UNDERSTAND ME?

Interrogator (*now talking softer*): Do you understand me? What are you carrying escape and evasion equipment for? Why are you carrying a machete? What are you going to do with that? What are you going to do with a machete and $150? You going somewhere nice? You're going nowhere nice. You're staying here for a long, long time . . . You're not prepared to help yourself. And that's a shame. That is a fucking long hard shame.

The training video demonstrated the abusive approach which, according to the officer at Chicksands, would be a precursor to 'normal' interviewing. But how easy would it be for a soldier to understand when 'harshing' stopped and inhuman treatment began? Would all the reminders about the Geneva Conventions really help them appreciate the difference? How would they know in the torrid environment of a war against an unseen enemy in a distant land that shouting abuse was acceptable, but any other cruel handling wasn't?

These questions were unanswered. Nor was it explained what the 'shock of capture' meant for those interrogators sent out to prepare prisoners for questioning. The Chicksands officer had been adamant that interrogators were not trained to maintain the 'shock of capture' through violence. He had said that the prisoner might experience 'pressures' both self- and system-induced. What did this mean? And, again, how would it have been taught?

One of the documents handed over to Gordon was headed 'Pressures on Prisoners and Detainees'. On the face of it the document appeared reasonably innocent. It emphasised that British interrogators were constrained by the Geneva Conventions and the law of armed conflict. They weren't supposed to induce 'mental or physical pressure in order to break a prisoner's will to resist'. It then went on to say that 'system induced pressures that are either unavoidable or that the capturing force may apply whilst complying' with the law will be allowed. The document listed these types of pressures in a seemingly benign fashion:

Dislocation of expectations	Unaccustomed discipline
Prison diet	Enforced idleness
Confinement and lack of comfort	Atmosphere of ruthless efficiency
Lack of news	Loss of companionship
Loss of sensory contact	
with the outside world	Mental fatigue
Lack of sleep	Mistrust of comrades

Feeling of failure	Sexual frustration
Inferiority complex	Show of knowledge
Interrogator's personality	Transference

Many of these items were pregnant with violence. Lack of comfort, loss of sensory contact, lack of sleep, atmosphere of ruthless efficiency, fatigue, sexual frustration. These were all permissible. But how easy would it have been to distinguish between what was allowed and what would contravene the Geneva Conventions? If 'lack of sleep' was tolerated then would dropping an iron bar on the floor to keep detainees awake be OK? If fatigue was acceptable, then would making prisoners stay in awkward body positions be allowed? And how would sexual frustration be systemically induced? These guidelines were supposed to be handed over to interrogators who might have a prisoner for maybe a few weeks at most. How would they create sexual frustration in such circumstances? How would an interrogator interpret such a possibility? If these 'pressures' were allowable as part of a system of capture and detention, then how would a soldier be capable of drawing lines?

The document's caveat that any pressure would have to be assessed in accordance with the law of armed conflict may have looked fine, much like those small-print disavowals of responsibility in insurance policies. In the world of a dirty prison in a conflict zone where officers are baying for information, would an interrogator know how far to go? Were the lines sharp and bright?

4 2

*T*ORTURE. *IT WAS ONLY A word. A word packed with emotion and political fright. Was it appropriate for the conditions and treatment of the men detained from the Hotel al-Haitham? WO Spence couldn't have chosen a worse time to consider the possibility. The Guantanamo Bay saga had begun to filter into public consciousness and the United States administration's discussions about what was permissible when interrogating those captured in Afghanistan and elsewhere had goaded outrage in the press. Donald Rumsfeld, the US Secretary of State for Defense, had authorised techniques that many legal experts and human-rights activists would denounce as clearly torture or inhuman treatment, both banned under international law. When internal documents were revealed under freedom of information requests in the US, the position of*

Rumsfeld and his legal advisors became open to ridicule. Not only were their attempts to justify methods such as waterboarding, solitary confinement and psychological terrorising wholly unconvincing, but they looked so forced. Why they chose to rationalise their interpretation of how detainees could be treated in law remains a mystery. The recommendations made, and the personalised margin notes signed by Rumsfeld, suggested the US administration condoned torture, practised torture and would most likely encourage its use by their allies.

WO Spence may not have known about the details of these memos or their implications for his investigation. His enquiry as to the interpretation of the word 'torture' at least implied that he was aware of the strength of the case emerging from his investigations. He sought legal advice. It isn't clear what he was told in response. But here is a good guess.

Torture now has political and social connotations that mean its allegation has to be carefully scrutinised. Inevitably, this involves sticking to those international agreements which have tried to outlaw torture. Perhaps the most persuasive of these is the United Nations Convention against Torture and other Cruel, Inhuman or Degrading Treatment or Punishment. It was produced in 1984 and represents the basic reference point for judging the activities of state authorities and was signed and ratified by Britain on 8 December 1988. Article 1 says that 'torture' means

> *any act by which severe pain or suffering, whether physical or mental, is intentionally inflicted on a person for such purposes as obtaining from him . . . information or a confession . . . when such pain or suffering is inflicted by or at the instigation of or with the consent or acquiescence of a public official or other person acting in an official capacity.*

If you unpick this long-winded statement, then a problem appears. It was leapt on by USA government advisors when considering the case of Guantanamo Bay. How do you define 'severe'? Exactly how much violence turns ill-treatment into torture?

This might sound like the proverbial angels dancing on a pin. Surely we know when pain is severe. But it's never quite as simple as that when it comes to the world of lawyers and the law. They operate in the realm of evidence and proof, of exactitude and certainty. If doubt is allowed to enter, then conviction (the very word suggests a belief founded on some certainty) can't be guaranteed. And any criminal prosecution in Britain has to abide by the 'beyond reasonable doubt' proof threshold. So, when a rule talks about 'severe' you can be sure that lawyers have to know precisely how they can show when pain or suffering reaches that level.

In relation to the techniques used on the detainees at Battle Group Main – the hooding, the plasticuffing, the sleep deprivation, the stress positions and the occasional kicks and punches – the lawyers already had something to help them answer WO Spence's query. The Northern Ireland case which had condemned these practices back in the 1970s was the source. When the judges at the European Court of Human Rights eventually looked in detail at what had happened in the prisons of Belfast and Derry they had decided that calling 'torture' the 'five techniques' of stress positions, sleep deprivation and the rest, which bore great similarity to the experiences of the detainees in Basra, was going a step too far. They were more comfortable with 'inhuman treatment'. The British Army lawyers didn't have to look much further than this. They didn't need to copy the Americans and say torture only related to the most extreme levels of violence. But on the strength of human-rights precedent, they could say that 'torture' would be an inappropriate description of the experiences of those detained by 1QLR. It was enough to call it 'inhuman treatment' as the European Court had done and, for the killing of Baha Mousa specifically, 'murder' would surely suffice. There would be no need to hinder any possible prosecution by having to prove a much harder case. Far better to avoid 'torture' as the focus of allegations.

All of this is conjecture, but it is more than plausible given that accusations of 'torture' didn't surface again in any meaningful way in the official investigation. That this relieved some of the pressure on the higher command levels may have been fortunate for them. Certainly, official recognition that torture may have been practised would have sent some damning signals to the general public. Those in the army and government would have imagined the headlines: 'Torture practised by the British Army in Iraq, say prosecutors' or 'British officers oversee regime of torture in Basra' or 'Soldiers charged with torture of Iraqi civilians'. It wouldn't have looked good, particularly given the widespread opposition to the Iraq War and the occupation in general; it would have been one more damning indication that Britain had breached fundamental rules of international law.

That such headlines would be written in any case couldn't have been foreseen with any certainty. At the time the lawyers would have advised according to what they knew of the law. And the law was vague enough to inspire caution. From this point it was decided that no one was to be prosecuted for torture.

43

O N 24 October 2003 WO Spence packed up the Baha Mousa case file and exhibits. They were placed in boxes and taken to Basra International Airport, loaded on to a Boeing C-17 transporter plane and flown back to SIB headquarters in Britain.

Members of 1QLR also packed up their equipment and readied themselves to vacate Battle Group Main. Officers and men had finished their work in Iraq. A few months in one of the most dangerous places on earth and chased out with investigations and allegations hanging over many in the regiment. There must have been very mixed feelings about the end of their mission. Some were proud to have fought and come through unharmed, at least physically. They would have stories to last them years of reunions. There must have been a sense of foreboding too. The weeks of interviews and questions and RMP officers prying into their lives in Basra might be drawing to a close, but they were unresolved. Sure, some of them had been plucked out of their midst without warning and with frightening speed, disappeared almost. They were first in line. But would that be the end? Surely they couldn't believe that. They weren't that naïve. They would have known that returning to Catterick Barracks would not protect them from scrutiny. It can't have been an appetising prospect for any of them, officers or men.

WO Spence followed the case documents and 1QLR home. Captain Gale Nugent, who had overseen the investigation from its inception, took the same flight back to Britain. Their tours of duty in Iraq had ended. The investigation had hardly begun.

Part 2

_____

IN BRITAIN

*B*ULFORD COURT MARTIAL CENTRE ON *Salisbury Plain has only just been completed and handed over to the army by the building contractors. Millions of pounds have been spent on its design and embedded technology. There are computer terminals on each desk, a sound recording system, comfortable executive chairs. Musty-smelling solemnity replaced by modern efficiency. It's 19 September 2006 and the first case in the plush new chamber is about to be heard. Almost exactly three years after the Royal Military Police began their investigation into the killing of Baha Mousa, the man accused of causing his death and six other soldiers are escorted into the trial chamber.*

The hearing has been going on for over a week now but the accused have been absent and the theatrics of advocacy largely suppressed. Something akin to a sophisticated traders' market has been taking place with the ranks of experienced QCs and leading junior barristers arguing over the minutiae of procedure and protocol. How should the press be allowed to report proceedings? What pictures of the defendants can they print? How should sensitive military information be presented? Days of these legal arguments have induced a state of flaccid hypnosis for the bevy of journalists and clerks and bag carriers. Many have retreated outside to experience the autumnal sunshine and have a smoke in the calming Japanese gardens in the quadrangle outside the chamber. But any tranquillity is more likely to be the result of tedium than enlightenment.

With the seven accused taking their seats, the trial proper begins. Julian Bevan QC, a rather creased Old Etonian who has spent over forty years at the criminal bar, is the leading barrister for the prosecution. He's been working hard with the Army Prosecuting Authority to construct a case against the defendants. The evidence accumulated by the Special Investigation Branch of the RMP is voluminous. There are albums of photographs, diagrams, army orders, logs, interview transcripts, statements (typed and handwritten), videos (of the detention centre where Baha Mousa was killed, of his body lying in the morgue, of the post-mortem), doctors' reports, bundles of correspondence. Boxes and boxes of material have been collated, indexed, paginated, copied. It has been a vast enterprise of logistics and intense scrutiny to make it ready. Teams of clerks charged with handling the documentary piles transport them in and out of the courtroom on little trolleys. Despite the scale of documentation, Bevan is its master, or so he would like to appear. For this is one of those historic cases where reputations can be consolidated or shattered. It's no ordinary criminal proceeding. It's a war-crimes trial.

The significance is not lost on the British media. Journalists from every national newspaper and TV news station have been anticipating the story for months. Not only are serving soldiers being prosecuted for offences inextricably associated with those criminal tribunals at Nuremberg and The Hague, but it also speaks of a political decay in the government of Tony Blair. Amidst a succession of critical stories surrounding the preparation for war against Iraq, its dubious legality and the botched handling of the country's occupation, the trial at Bulford fits neatly into a common theme, one to be quarried mercilessly.

Of course, Julian Bevan makes no reference to wider political matters when he opens the case. That would be to introduce a piece of grit into a system which cannot tolerate tarnish. Bevan's duty is to keep rigorously to the charges before the court. He mustn't allow sentiment, political or otherwise, to interfere with the simple working of the law.

He begins, addressing the judge advocate, Mr Justice McKinnon, who is the first civilian judge to preside over a court martial, and a jury panel comprised of a major general, two brigadiers and two full colonels.

'I am going to take you back to September 2003,' he says, 'almost exactly three years ago. In September of 2003 a number of Iraqi civilians were arrested in Basra, Iraq, following a raid on a hotel by soldiers of 1 Queen's Lancashire Regiment. They were taken, having been arrested, to the Battle Group Main headquarters and they were there detained in a temporary detention facility. The reason they were taken to that temporary detention facility was for a decision to be made as to whether they should be interned on the basis that they posed a threat to the Coalition forces. Those Iraqi civilians were in fact detained in this temporary detention facility for about forty-eight hours before being

taken to the theatre internment facility, commonly called the TIF. This case centres upon the treatment of those Iraqi civilians when held in that temporary detention facility. It is the Crown's case that over a period of about thirty-six hours, extending from the morning of Sunday 14 September certainly to late evening of Monday 15 September, whilst in that temporary detention facility, which as you will see is a small three-roomed building without doors, very near in fact the centre of operations, whilst in that building, that temporary detention facility, they were repeatedly beaten, says the Crown, by being kicked and punched at a time when they were handcuffed, when they were hooded with hessian sacks, made to maintain a stress position for unacceptable lengths of time, deprived of sleep, continually shouted at and generally abused in temperatures rising to almost sixty degrees centigrade. That, as you all will appreciate gentlemen, is well over a hundred degrees Fahrenheit.'

Bevan soberly lays out the charges. There are no histrionics, no tub-thumping speeches about the evil men do, no angry condemnation, just simple description. One man is accused of three crimes: manslaughter, inhuman treatment and perverting the course of justice. Two are accused of inhuman treatment alone. One more is charged with simple assault. And three, two of whom are officers, with negligently performing a duty.

Bevan sets about explaining the evidence the panel will hear, the choices they will have to make about guilt or innocence, and the difficulties they will face in trying to reach their verdicts. And he takes his time about it. For three days he methodically animates the scene of the killing. He describes the location, the legal framework, the people who will appear to give evidence, and the way in which he hopes the trial will proceed. He speaks to the panel of military men with the intention of convincing them that this is a serious case, demanding an intense commitment from each of them. He explains its importance and tailors his language in such a way that anyone might understand the criminal nature of the accused soldiers' behaviour.

But scan the benches and seats and the corridors outside and there's one person missing. Daoud Mousa is not here despite the explicit promise that he would be invited to any court martial relating to the death of his son. The father is left at home in Iraq to lament the son without even the satisfaction of seeing those accused confronted with their crimes. It's as though the abuse has no end. Shards of broken pledges continue to pierce the immaculate carapace of British justice.

Bevan isn't interested in the father. He's not that concerned with the son.

His attention is focused on the seven accused, what they did or didn't do, the nature of the charges against them, and how he is going to prove their guilt beyond reasonable doubt. But the success of his endeavour will hinge on more than his arguments. The investigation which continued when protagonists and detectives returned to the UK explains much about the fate of Mr Bevan's case for the prosecution.

1

WHEN WO SPENCE RETURNED FROM Iraq in October 2003 he couldn't have imagined his investigation of the Baha Mousa case would take a further three years to conclude. Not that any particular urgency was evident from the case file. The move back to the UK was an upheaval in itself, bound to affect the good order of any inquiry. It was chaos. And then Christmas soon intervened. No progress was recorded between 18 December 2003 and 6 January 2004. Even in the New Year there was little discernible change in the speed of the inquiry. Attention was then focused on arranging identification parades. As WO Spence was refused permission to bring the surviving Iraqi detainees over to the UK and the soldiers under suspicion couldn't be transported back to Basra from Catterick (where they were now based) he had to video the suspects and take his recording out to Iraq. It wasn't entirely satisfactory but it would have to do. Late in January 2004 Spence made the trip armed with the DVDs and laid on a film show for the detainees. Positive IDs of two of the suspects was all he could obtain.

His inquiries became even more complicated on his return. 1QLR was due to depart for Cyprus on 21 February. Now he would have to coordinate inquiries across three countries. But there was one advantage in the regiment leaving Britain again. It would take it away from the growing glare of press attention. Even the *Sun* was showing an interest. The Ministry of Defence had been able to bat away the rumours of ill-treatment and unlawful killing by assuring everyone that a thorough investigation was in hand (those had been the 'press lines' since the previous October), but if more stories started to appear in the tabloids there was no knowing what would happen.

Spence prepared to follow 1QLR to Cyprus to re-interview the men arrested but not yet charged: Corporal Donald Payne, Private Craig Slicker and Private Peter Bentham. He needed to unpick their stories and test their evidence. In the early part of March 2004 he flew to the British base in Dhekelia, southern Cyprus.

The island was not unfamiliar with this type of investigation. In the 1950s shocking stories of crude brutality were commonplace. The British Army had been there trying to deal with an insurgency against colonial rule. Again the problem was intelligence. And the solution was a system of interrogation of suspects directly ignoring the terms of the European Convention on Human Rights which the Foreign Office had had such an instrumental hand in drafting. Detention was introduced under Emergency laws in 1955, insurgents were labelled 'terrorists' (and indeed many terrible atrocities were committed by those fighting the British) and suspects were allegedly subjected to torture and ill-treatment. The Greek government objected vehemently. Its public outbursts prompted the British authorities to undertake official inquiries. These revealed that two British officers had been court-martialled for ill-treating suspects. The Greeks suggested this was insufficient and publicised further allegations which were specific and lurid: sleep deprivation, a walking stick thrust up one victim's rectum, beatings with straps, and variant forms of water torture. There were two deaths in custody among the complaints. One had been 'trying to escape' and had supposedly struck his head on a rock after being rugby tackled by a British officer. These weren't the only occasions when violence had been inflicted on Cypriots taken into custody. British soldiers spoke later of brutality used against the local population, admitting their crude contempt for their enemies. It was just another of those dirty little wars where the treatment of detainees always seemed to provide an uncontrolled point of violence between occupier and suspected insurgent. Raw disdain for the 'enemy' spilled over into crude barbarity.

Over a few days on the island, WO Spence conducted his interviews. No one could claim it was a great success. Hardly anything was added to the picture already formed. The responses to the questions elicited little that wasn't known. Corporal Donald Payne stuck closely to his story. He had been ordered to carry out the conditioning of the detainees, he said. He was in no position to question those orders. As far as he knew, all the officers understood what was going on. Major Peebles had sanctioned the methods. Payne denied assaulting any of the detainees. There had been no choir, no punches, no visitors who could have seen him attack anyone. And when it came to the death, Baha Mousa was trying to escape. That's what he believed and that was why he had tackled him. It was his job, his duty. He *had* to restrain him. The prisoner was resisting. What else was he supposed to do? Others had helped him, but the man was struggling and probably hit his head in the struggle. They had tried to save his life, but resuscitation hadn't worked. And they had called the medics immediately. These were hardly the actions of soldiers out to kill. And no, he hadn't kicked or punched Baha Mousa during all this. He was just trying to get him back under control.

None of the other men interviewed revealed anything new either. WO Spence had to return to Britain with all his lines of inquiry still in their infancy.

Unknown to him, however, some blockage had been punctured once the regiment had left Iraq and relocated to Cyprus. Perhaps the mournful inaction on an island of quiet guard duty and meaningless stock-taking had induced deeper reflection. Or perhaps Spence's questioning had provoked some consciences. Within a week of Spence's visit to Dhekelia a member of 1QLR came forward. The soldier was Private Jonathan Lee. He was one of the Hollender multiple of A Company in Basra, who had been largely left alone by the investigators in Iraq. In the middle of March 2004 he sought out the regimental padre in Dhekelia with a plea more than a confession. He wanted to report something he had witnessed in Iraq, he said. The padre advised him to go to the military police. On 18 March he contacted Warrant Officer Thompson, one of the RMP officers within the regiment. They sat down together and Private Lee began to unburden himself. He was a worried man. He said soldiers had begun to suspect he was going to talk to the SIB. He felt threatened. There had been hints, aggressive glances. He wanted protection. He would say that he was near the detention centre soon after the detainees were first delivered to Battle Group Main. He would say he went inside, intrigued by sounds of shouting and screaming. He would say he saw Corporal Stacey and privates Crowcroft and Fallon kick and punch the prisoners. He could remember specifics and he would testify in court if they wanted. But they had to protect him.

2

AT THE BEGINNING OF APRIL 2004 Captain Lord of the SIB reviewed WO Spence's case file on the Baha Mousa killing. Lord reported that 'a thoroughly professional inquiry meticulously maintained and a credit to the warrant officer in charge and his team' had been conducted. There were lines of inquiry outstanding, but in his opinion all was going well.

Within three weeks the 'shit had hit the fan' as some in A Company would describe it. The whole investigation came under minute and very public scrutiny. Britain's media were itching for sordid tales of wrongdoings in Iraq. And by a very convoluted route, the story of Baha Mousa and his fellow detainees had begun to capture the interest of journalists.

When the British press get hold of a story they are dogged in its pursuit. All sorts of methods of uncovering information are used, some not as legal as others. In May 2004 they hardly needed to lift their collective heads to be drenched by waves of revelations about Coalition forces and their criminal behaviour in Iraq.

The first wave came from the USA. Rumours about some shocking exposure of prison conditions in Iraq had been filtering through media channels for a few weeks, but it was a TV news report which heralded uproar. The 28 April 2004 edition of the programme *60 Minutes* on CBS News broadcast a slideshow of photographs of US troops humiliating and abusing Iraqi men inside a prison facility. A female guard pointing at the genitals of naked inmates; the stripped bodies of several men piled high in a corridor, two soldiers grinning behind the bizarre tableau; the same female guard holding a dog leash attached to a collar worn again by a naked man lying on a concrete floor. The images were extraordinary. They possessed a wicked intensity worthy of a Pieter Bruegel painting except that the smiling guards made the viewer think they must be just a joke. There was nothing funny for those subjected to the humiliation and, it would later appear, the torture.

After the programme came the press reports, the analysis, the condemnation. These photographs, apparently taken as debauched trophies, had been recorded on mobile phones by the guards themselves. They brought to international prominence the name 'Abu Ghraib', the prison where they had been taken. This had been one of Saddam Hussein's infamous jails. Rather than disband it after the invasion the US forces had adopted it as their own. They needed a place to process all those suspected loyalists to the Saddam regime and all those so-called potential terrorists. Now there was evidence that the Americans had adopted similar methods of treatment for its inmates to those used before the 'liberation'. The irony wasn't lost on many commentators as the US went into a period of self-loathing analysis.

The story and the pictures were reproduced in the UK. The British
media soaked up the emerging exposé. It followed a sustained critique of
the situation at Guantanamo Bay, which had started to outrage human-
rights activists with its legal 'black hole' status and suspected coercive
interrogation methods.

Then, almost miraculously, on 1 May 2004 the London press had its own
photographs. The *Daily Mirror* published a scoop of pictures allegedly taken
by British soldiers abusing an Iraqi civilian. They weren't as graphic as those
from the US, but they were just as effective in their portrayal of 'our boys'
disgracing themselves and the whole British Army. Piers Morgan, then editor
of the paper, vouched for their veracity. The pictures showed Iraqis in hoods
in the back of a truck being beaten and urinated on by British soldiers dressed
in desert combat gear. Two squaddies, named only Soldier A and Soldier B
by the paper, had purportedly taken the photographs and given their story
to a *Daily Mirror* journalist. Their identities weren't revealed. The news article
reported their accounts of violence done to Iraqis and accused members of
the Queen's Lancashire Regiment as the culprits. Such brutality was endemic,
it was claimed; hardly conducive to winning over hearts and minds.

The paper reported the soldiers' story of a civilian picked up in Basra
and subjected to a horrific attack by them and their mates. Soldier A
claimed that the man was given a beating with batons. He said there was
blood and spew and urine everywhere. The victim's jaw was 'out'. He had
been thrown from the back of the truck, left with his injuries. They didn't
know whether he lived or died.

The whole story was lurid, but it was the photographs that lent it its
power. That soldiers would act like this was bad enough, but to take
pictures of it as well simply accentuated the moral catastrophe for the
British Army. The political reaction was prompt. Armed Forces minister
Adam Ingram said the photographs were 'deeply disturbing'. There would
be a full investigation, he assured Parliament. General Sir Michael Jackson
said the soldiers involved weren't 'fit to wear the Queen's uniform', but
pleaded that the whole army shouldn't be judged by a few criminals within
its ranks. It maintained the highest standards of behaviour, punished those
who transgressed, and it would act swiftly in this case to identify and
prosecute any culprits, Jackson said.

The general's comments were applauded by the *Daily Mirror*. The paper
began to run a parallel story to the torture claims: the allegations should

be investigated, the paper reported, but that should not detract from the heroic conduct of the vast majority of those serving in Iraq. A distinction was made with the American forces. In contrast to them, so it was implied, the British were humanitarian, professional and courageous. The paper's editorial on 1 May 2004 described the troops as renowned for their 'discipline and self-control'. The 'sick minority' should be punished quickly and the paper was heartened to hear General Jackson's emotional promise to ensure that that would happen. But it warned that none of this should detract from the heroism that typified the 'best army in the world'.

The *Mirror*'s patriotic message took firm root in the newspapers shortly thereafter. It would last well beyond Iraq and underscore all reporting on the British military presence anywhere in the world, particularly in Afghanistan. For a time there would be a conflict between the condemnatory tone directed at British forces for their part in harming innocent civilians in far-off wars and the ingrained story that the troops were fundamentally heroic. The latter triumphed, attaining mythic status, but the battle between the two lasted for quite some time. Neither side profited from the *Daily Mirror*'s supposed revelations.

Within a couple of days of publication, the photographs – not the abuse they portrayed – became *the* story. The BBC revealed the next day that the pictures had been faked. Ministry of Defence experts and members of the Queen's Lancashire Regiment had examined the photographs with forensic efficiency. They had spotted several details which didn't add up. The soldiers were wearing the wrong type of hats (floppy instead of berets or helmets), they were carrying rifles which had never been issued to troops in Iraq, the stream of urine 'didn't look authentic', army boots were laced in a criss-cross fashion when the army preferred them in parallel. And, with a final flourish, the Royal Military Police had matched distinctive scratches and marks on the truck in the photographs to one 'found' in a Territorial Army barracks in Preston. The analysis was pedantic, but damning. Whoever had undertaken the investigation had acted with extraordinary speed. How they had managed to track down the exact same truck in twenty-four hours, carry out the examination and confirm the match wasn't publicised. It was an investigatory miracle.

Despite standing by its story for a few days, the *Daily Mirror* eventually had to accept that the photographs they had bought were faked, mock-ups. Piers Morgan was dismissed from his post as editor. An apology was printed: 'Sorry . . . we were hoaxed' was the headline. The paper presented itself as a victim of deception. It was almost a great press tradition: journalists suckered into believing they were being handed a fabulous scoop whereas in fact all they had was an amateurish fabrication. People still remembered the Hitler's diaries debacle in the early 1980s when *Stern* magazine and the *Sunday Times* were hoodwinked along with a notable British historian. It had been a lesson for editors which the *Daily Mirror* had perhaps forgotten. Now it had to backtrack.

Geoff Hoon, the Secretary of State for Defence, and Brigadier Geoff Sheldon of the Queen's Lancashire Regiment both welcomed the retraction. There was an air of smugness about their responses. It was 'good to know the *Daily Mirror* had done the right thing so quickly' was the message; the hoax was malicious but the paper's donation of money to an army charity was accepted with magnanimous approval. Honour had been restored to the maligned regiment.

On the same day as the *Mirror* published its admission, the Royal Military Police confirmed that they had arrested four men over the Baha Mousa allegations. The news was drowned by the furore and incestuous press commentary over Piers Morgan's dismissal. No one seemed to care that although the pictures were fakes, the abuses weren't.

3

I F THE BAD PUBLICITY PROVOKED by the *Daily Mirror* photographs had been dampened by swift action (in marked contrast to the ponderous investigations in the Baha Mousa killing), the government and army weren't so lucky when it came to other stories emerging.

On 4 May 2004, as the *Mirror* fought to authenticate their pictures, the *Wall Street Journal* published a leaked confidential report by the International Committee of the Red Cross. Although US transgressions were the focus, UK forces weren't immune from criticism. The report had drawn on the Baha Mousa case as a specific example of general allegations against British units. It also had much to say about the treatment of prisoners at Camp Bucca. Interrogation methods were criticised for their brutality. The report listed the offending actions in clinical fashion. At the head came the practices of hooding and handcuffing with 'flexicuffs', but the systematic use of threats (a favourite being transfer to Guantanamo Bay, as though that had now become the watchword for the extinction of hope) and ritual humiliation (usually through rendering the prisoner naked for long periods of time and in open view) were all cited as unacceptable.

The complaints were so well defined and so reminiscent of the experiences of the detainees of 1QLR that British officials must have shuddered. Following so closely the pictures of the abuse in Abu Ghraib would only have increased the torment. All those press expressions of outrage at the conduct of US personnel could now just as legitimately be turned upon the British. There was no distinction between the two armies in the view of the ICRC report. These were 'Coalition forces', neither American nor British. They acted in unison, shared the same camps and it would seem the same methods of detention and interrogation and ill-treatment. The report concluded that 'serious violations of International Humanitarian Law' had occurred. Prisoners faced the 'risk of being subjected to a process of physical and psychological coercion, in some cases tantamount to torture, in the early stages of the internment process'.

The very fact that the ICRC report had found its way into the public domain was a story in itself. Unsurprisingly perhaps, the belief that the ICRC had leaked the story as a means of criticising two member states of the UN Security Council began to take hold. There was a suspicion that political motives were at work. Indeed they were, although it later transpired that a member of the US administration was the informant and not the ICRC. Nonetheless, that the Red Cross should be condemning these pillars of the international community even in secret was a political disaster. It gave added weight to the anti-Iraq War movement; not only were the British and USA governments dismissive of the relevance of international law in going to war, but they also abused the basic principles that they proclaimed were a reason for going to war with the Saddam regime in the first place. The protection of human rights was a central plank of the rhetoric surrounding the invasion. Overthrowing an abusive regime was demonstrably a 'good thing' so the public were repeatedly told. This ICRC report suggested one violator was being replaced by another.

Baha Mousa was to come to personify the allegations and provide the evidence that British forces were indeed in contravention of proclaimed national values and standards. It was an indication that abuse was endemic, the culture of the armed forces serving in Iraq. Could this really be true? Was the British Army so infected?

4

DESPITE THE *DAILY MIRROR* PICTURE hoax, the stories of maltreatment wouldn't disappear. The day after the paper made its apologies, the ITV programme *Tonight with Trevor McDonald* featured another exposé. An unnamed soldier from the Queen's Lancashire Regiment wanted to tell his story to the world. Or so the programme advertised. His identity was disguised because he feared for his safety. He was able to say how detainees had been treated with contempt and violence. It sickened him that army spokesmen would come on TV and deny that these abuses happened regularly. It made him angry. He had come forward because those in command were simply hoping the stories would go away, that they would be buried. As far as he was concerned, they *knew* what had been going on.

The edited conversation between Sir Trevor McDonald and the anonymous soldier made for extraordinary television.

The soldier said 'I just wish that instead of trying to dismiss everything they should hold up their hands, take some responsibility for some of it.'

McDonald said 'What did you see?'

The soldier said 'I saw prisoners punched, slapped, kicked, pushed around. Sandbags zipped tight. I saw them in those sandbags for hours and hours on end.'

McDonald said 'Were they beaten?'

The soldier said 'Yes, they were beaten. They were beaten for fun.'

The soldier condemned the lack of control exercised in Iraq. He had seen things that deeply disturbed him. No one had said anything at the time to stop it.

The soldier said 'It was not all the army. It was not systematic. But it did happen. It was isolated incidences and I believe a lot of the British soldiers didn't know that it was going on – maybe that's part of the problem, that it's kept so hush-hush when things are reported.'

Lance Corporal Ali Aktash was the soldier who gave the interview. A member of the Territorial Army, he had been mobilised in July 2003 to serve in Iraq. He was trained as a signaller and was assigned to 209 Signal

Squadron initially. Halfway though his tour, though, he was transferred to 1QLR. On 15 September he was working in the operations room at Battle Group Main. At some point he went to the detention block to use the Portaloos outside. He heard noises coming from within the building; the sound of hitting, thumps. Like so many other soldiers he had walked into the block. He saw Corporal Payne slap and shout at the detainees and at one point put his hand around the bagged head of a prisoner and force his thumb into the man's eye. He saw other soldiers abuse the detainees. Aktash said that prior to being taken for questioning, a couple of guards made a prisoner run around outside the detention block. As he had no belt, the prisoner's trousers kept falling down. Everyone laughed and mocked. 'It was quite a comical sight,' Aktash admitted. But the acts of random viciousness rather than idle humiliation shocked him.

Even before Trevor McDonald's programme got hold of his story Ali Aktash had revealed what he knew to the SIB. He had been caught up in the *Daily Mirror* photograph story. It was this that hauled him into the media tornado.

He had finished his tour of duty and was back in the UK when the *Mirror* pictures appeared. A friend had spoken to him in the pub about the pictures. Aktash had told his mate how they could well be true – that kind of stuff did happen, he had said. Somehow the information had been passed to the *Mirror* who called Aktash when suspicions about the validity of the pictures began to appear in the press. He was persuaded to go the *Mirror*'s offices in London and look at the photos. Perhaps it was thought he could vindicate the paper's decision to publish the pictures. It was a rearguard action that eventually proved fruitless. But it also brought Ali Aktash to light as another witness of fact. He was one more for the SIB to add to the growing list.

5

6 MAY 2004 WAS A day of shame for British justice, according to the *Daily Mail*. On that day the paper published a commentary on the Stephen Lawrence affair. Mr Lawrence was a black teenager who had been killed by a gang of white racists on the streets of London. Police incompetence had contributed to his killers escaping trial or conviction. The evidence against the white youths was simply insufficient, prosecutors said. Failures in the investigation and the handling of the matter by the

Metropolitan Police prevented a case being brought. The paper accused five men of Mr Lawrence's murder. It challenged them to sue for libel if they dared.

On the same page as this provocative piece there appeared another article on the subject of British justice. 'Now comes the final insult to our fighting forces', it proclaimed. The target of its ire was the statement by a lawyer that he was acting for Iraqi families who had lost relatives allegedly killed by British soldiers after the war had ended. Extraordinary vitriol was directed at this lawyer. He was 'publicity-grubbing', 'poison-tongued', 'in search of a fat fee'. His announcement that an action was being lodged with the courts to seek a proper investigation into the deaths of twelve Iraqis under the terms of the Human Rights Act was seen as a betrayal, a mark of the compensation culture. That Iraqis should gain any benefit from human-rights standards was 'lunacy'. The paper's ongoing opposition to the introduction of the Human Rights Act may have underpinned the piece but the attack against the lawyer who had used the word 'murder' to describe one of the killings was the focus of its outrage.

Phil Shiner was the lawyer in question. The day before the *Mail*'s article was published he appeared on the *Today* programme, the most influential news broadcast on BBC radio, to talk about the cases he was bringing on behalf of Iraqi clients. His allegation of 'murder' was said in the context of one of the twelve matters he was pursuing. During the interview he talked about this one case as being particularly disturbing. Although most of the others involved the alleged unlawful killing of Iraqi citizens, it was the death of Baha Mousa which he described as an act of 'murder'. This killing had taken place within a British Army base, he said, the other deaths had occurred on the streets of Basra or in private homes during army raids. It was an important distinction.

Although the difference would have mattered little to the victims and their families, this set the Mousa case apart. It prompted the question: would British law apply to a British camp located outside the UK and in the sovereign territory of another state? If it did, then the human-rights standards that people on British soil could expect, specifically the right to life, would apply. The family of someone killed by British forces would then be entitled to a full and proper investigation into their relative's death. This hadn't happened in the Baha Mousa case, or so it was argued.

That was the gist of the claim. It was revolutionary, as far as any legal action can truly be described as revolutionary. The Human Rights Act remained a new and relatively untested piece of legislation. No one was quite sure how it should be applied in the labyrinth of circumstances that might arise.

Whether the killing amounted to murder was another question, but it was a reasonable accusation to make given the facts known at the time.

The *Daily Mail* didn't agree. Max Hastings, a distinguished journalist and author of popular military history, was brought in to comment. He lambasted the idea that British troops could be expected to watch out for every little human-rights issue whilst patrolling the streets of a battle zone. It was utter madness, politically correct madness, he wrote. And New Labour had brought it upon themselves with their liberal zealotry, the paper said. But the legal claim had been lodged in the High Court and the action would continue whatever the *Daily Mail* or Max Hastings thought. They couldn't interfere with the due process of law. Nor could they prevent the lawyer from carrying out his duty to represent his clients to his utmost.

6

Iт's hard to think of any other professionals who attract the kind of jokes that lawyers do. Here's one of the milder ones.

There's an island in the middle of the ocean. A doctor, a dentist and a lawyer are marooned there. They see an abandoned lifeboat floating beyond the island's reef and out to sea. One of them must swim to it and bring it back to the island so that they can escape to civilisation. But the island is surrounded by sharks. The doctor and dentist say they can't swim. Seeing through their lie, the lawyer says 'I'll go', and he dives into the water. Instantly, the sharks part before him and the lawyer swims unharmed to the lifeboat and returns with it to the island. The doctor and dentist look at the lawyer in awe. 'How did you manage that? Why did the sharks let you pass?' asks the dentist. The lawyer smiles and replies 'Professional courtesy, my dear chap, professional courtesy.'

Lawyers don't generally have a good reputation in Britain, at least not in the public imagination. The vitriol directed at the legal profession may not be quite as bad as in the USA, but sometimes it comes close. Occasionally movies or TV shows depict the crusading lawyer as a heroic figure. Gareth

Peirce, a British solicitor, achieved sympathetic fame when she was played by the actress Emma Thompson in *In the Name of the Father* about the fight against the miscarriage of justice of several people falsely accused of IRA bombings in the 1970s. But there are as many, if not more, portrayals of the devious and shady lawyer, the lawyer willing to compromise any ethical code for the sake of winning a case or pocketing a large fee. The philosopher Aaron Ben-Ze'ev, perhaps unwittingly, perhaps not, evoked much of the revulsion for the profession when he wrote about 'moral disgust'. He gave three examples: 'Nazis, people who steal from beggars, and lawyers who chase ambulances to acquire new clients.'* Another writer, an American law professor called William Miller, called lawyers 'moral menials', akin to people in the 'system' who collect garbage or butcher meat. They exhibit the vices of hypocrisy or betrayal or even cruelty. They are contemptible *because* they work with 'moral dirt'.†

Media stories often play on these negative images and the perceived popular disgruntlement directed at the profession in general. A sort of phantasm is adopted, a collectively induced thought which is difficult to shake off. Maybe it has something to do with the association between the lawyer and trauma. When else are they needed except during rage-fuelled litigation or stressed house sales? And so the jokes about greedy, duplicitous, self-seeking, unethical, unsympathetic lawyers feed off and repeat the image in an ever-replicating cycle.

Some of the more hysterical newspapers like to repeat this offensive stereotype (or perhaps, more appropriately, 'trope') whenever they get the chance. 'Fatcat lawyer' is a common headline, applied with particular relish to those paid by legal aid, but invariably used whenever a well-heeled solicitor or barrister appears in the news.

Phil Shiner earns his living mostly from legal aid. His practice, Public Interest Lawyers, based in Birmingham, is largely funded by legal aid. It has been awarded a Legal Services contract to practise in public law matters. That is what it does on behalf of the legal system, along with many other firms of solicitors. The cases he brought on behalf of Daoud Mousa and the detainees at Battle Group Main were all officially approved for funding by the body dispensing legal aid. Did that make him a 'fatcat lawyer'?

---

* See Aaron Ben-Ze'ev, *The Subtlety of Emotions* (MIT Press, 2001).
† W. I. Miller, *The Anatomy of Disgust* (Harvard University Press, 1997).

In the eyes of the *Daily Mail*, there was no doubt that it did. But there are few signs of wealth and ostentation at Shiner's offices in the old and rather run-down Jewellery Quarter of Birmingham. There are no expensive brand-name shops here, no coffee-chain outlets or chic bars. Turning into the street where Shiner's office is located, the sight is of a row of dour and dishevelled buildings, anonymous and seedy, most in need of refurbishment. Scaffolding appears regularly along the street accompanied by the ubiquitous skip. One of the converted properties in this street houses the legal practice of the small team of lawyers over which Shiner presides.

If the outside is austere, what about the inside? There's no lavish wealth evident here either. The office is remarkably quiet, the phone ringing only infrequently. All the walls are full of shelving carrying meticulously ordered files and neatly printed papers. The order and attention to detail is profound. Everything is labelled in clear print; documents are numbered and tagged and filed with almost religious concentration. It's hard to believe any important paper would be lost amidst this shrine to order. But there's nothing brazen, nothing extraneous to the work that might be called a 'luxury', a trapping of the 'fatcat'. If anything, it has a monastic smell and feel about it; an ascetic environment where work is the devotion.

A few minutes spent in these surroundings confirms that everyone there knows very well who has the ultimate responsibility for all that happens within this building. Perhaps it's this individual burden which explains the absolute necessity for maintaining professional and meticulous order. It's a matter of survival. For by now Shiner knows that if he or his colleagues make mistakes, they will attract the attention of a merciless media. It's become that personal since he brought the Baha Mousa case to court.

If none of this reflects the 'fatcat' image now, what was it like in early 2004?

When Shiner first received instructions from Baha Mousa's father he was based in an even smaller, more unadorned and less salubrious office. It was on the top floor of a converted warehouse on one side of St Paul's Square in Birmingham. He was renting one big loft room that was served by a kitchenette in the corner and a small toilet outside. The chairs were uncomfortable and cheap. He was a sole practitioner, with only a secretary and an occasional assistant to help him, struggling to develop his human rights and environment law work. The state of the office was of little concern to him. What mattered more was the pursuit of the kind of cases which matched his convictions.

If all this was true, why was Shiner targeted as the epitome of all that was bad about lawyers? What made him such a figure of opprobrium for the *Daily Mail*? Perhaps he embodied the paper's hatred for supposed hypocrisy. Perhaps the *Mail* couldn't reconcile its perception of those legal beneficiaries of the years of City expansion and multimillion pound corporate deals with that of a radical campaigning lawyer. Perhaps he embodied for the paper the betrayal of 'our boys' sent out to fight a dirty little war with little preparation or protection and expected to behave with honour when faced by an unscrupulous enemy. Or perhaps the *Mail* just couldn't see beyond the simplistic myth of the 'shyster', the person who used the law for impure and self-satisfying reasons. Whatever the cause, the paper's portrayal helped construct a further barrier to finding the truth in the Baha Mousa case. It fouled the ground, at least for a time.

7

WOULD IT BE AN EXAGGERATION to claim that if Phil Shiner and his team of lawyers hadn't brought the action for judicial review of the investigation into Baha Mousa's death, the whole affair would have evaporated for lack of institutional interest and firm evidence? Would the army have sorted the mess out by itself, preserving its reputation for professionalism and leaving the vast majority of its troops untainted? Would the government have attempted to look deeper into the corrosion of values the affair suggested? It's doubtful.

The SIB's investigation was sluggish at the beginning of 2004. There were delays in completing witness statements, in obtaining identification of the accused, in bringing charges, in instigating proceedings. The very fact that the case eventually took three years to come before a court martial points to a certain laxity in prosecution. But it's hard to prove. It can only be speculation that the commencement of the High Court action by Shiner forced the government, the Ministry of Defence, the army and the Royal Military Police to take the case of the abuse of the detainees at Battle Group Main seriously. But the 'coincidence' of his litigation and an official seriousness applied to the case indicates some connection.

How, then, did Phil Shiner come to assume such a pivotal role?

Before the Iraq War, Shiner was deeply involved in the peace movement. He acted for the anti-nuclear weapons lobby and environmentalist campaigners. Law was a means to an end, a tool, sometimes to protect activists who chained themselves to the fences outside the nuclear weapons experimentation centre at Aldermaston, sometimes to spearhead outrage at government practices, sometimes to bring attention to a particular injustice.

When Tony Blair's administration threatened Iraq in the aftermath of the 9/11 attacks in the USA, few people in the peace movement doubted that some kind of conflict would take place. It was obvious from the beginning of 2002. Opposition to war quickly focused on the legality of any action against Saddam Hussein's regime. It wasn't that Saddam deserved protection. For years it was only human rights and peace activists who protested against atrocities in Iraq. But war as a means of punishment for something that the Iraqi people had had nothing to do with was abhorrent. There were better ways, they argued. It would be the civilians, the ones who had nothing to do with any weapons of mass destruction programme, which Saddam might or might not have had, who would suffer most. If any action was to be taken then it had to be lawfully approved by the United Nations, not a whim on the part of the US and its allies. That was the thrust of resistance.

During the latter half of 2002 the legal dimension began to assume increasing importance in the opposition to war. It was a rallying call. The argument came down to a simple formula: unless there was UN Security Council approval, invasion would be illegal, against international law. Everyone suddenly became an expert in this arcane subject, able to protest with this as the banner headline. Philip Allott, an eminent Cambridge University professor and an expert in international law, expressed bemusement because a taxi driver began to pronounce to *him* on the intricacies of legal process at the UN. You needed a UN Security Council resolution, the driver no doubt told the professor. The US couldn't just strike. It was wrong for Blair to say that this was self-defence. That was a distortion of decades of established doctrine. Everyone knew this now. Not just secluded academics.

Unlike the rest of the critical population, Phil Shiner wouldn't then claim expertise in international law. That didn't matter. The law could be learned. What did matter was how the legal process might have a role to play. It might allow for a challenge to what seemed to be an inexorable drift to war. Could law intervene to stop what everyone, from professor to taxi driver, was saying was illegal? That would make sense: using the law to enforce the law.

Of course, the system is never quite open to simple equations like this. As part of the peace movement, Shiner spoke with activists at the Campaign for Nuclear Disarmament in the summer of 2002 to explore how the law might have a role. CND had been a prominent organisation during the 1970s and 1980s, their protests set against the backdrop of the Cold War when the threat of oblivion seemed very real. After the Berlin Wall was taken down and Russia emerged as a putative western ally, their purpose diminished. They became a moribund organisation, barely registering with the public. Its long-serving members didn't disappear, though. Their credentials as peace activists were unchanged and when it became clear that war against Iraq was inevitable they sought to intervene. They instructed Shiner to bring an action in the High Court, a judicial review of the government's intention to go to war without a UN Security Council resolution. Predictably the case failed. The judges decided that this was one of those areas where the courts had no right to intervene. They had no power to require the government to take any particular action where foreign policy and military action were concerned.

Even though the case didn't succeed, the legal arguments against war remained. However, their centre of gravity in Britain shifted. The only court which could say the invasion was illegal now was the International Court of Justice, and that institution would never hear a case in time. Actions there could only be brought by another state and even then it would take years before the court would get round to listening to arguments and making a decision. For opponents to war this wasn't an option. But there were other pressure points.

Shiner and his colleagues began to think less about the legality of the war and instead considered *how* war would be fought. What would happen if British forces breached the laws of war? This was no longer an academic issue. In October 2001 the Blair government had ratified the International Criminal Court statute. Any war crimes or crimes against humanity or genocide committed by British forces could now be subject to investigation and prosecution by the International Criminal Court. It was a momentous change to the law. The chief prosecutor of the ICC had the power to investigate any serious international crime. Indictments of military and political leaders could follow. This was the first time that British troops would go to war with such an international body able to scrutinise their every action. It was the first time that generals and ministers could be held to account, at least in theory. And it was a theory that Shiner sought to test. If all those politicians and senior armed forces officers were put on notice that their actions might attract personal liability then maybe, just maybe, they might think again about embarking on war.

Bizarrely, it was a comedian who brought the threat of ICC scrutiny to greater public attention. Mark Thomas had already established a solid reputation as a radical stand-up comic. He was of that generation who had made a living out of protesting against the Thatcher government. The 1980s and 1990s had honed his art. By the beginning of 2003 when the issue of weapons of mass destruction was *the* inspiration for military action (or so it was claimed in London and Washington) Thomas produced a programme warning the British government about its responsibilities. He brought in Phil Shiner as his 'brief' and together they made an impromptu visit, on camera, to Downing Street to serve a letter on the prime minister. Although the intention was to introduce a note of ironic comedy, the letter was a serious legal threat. It notified the government that any breach of the laws of war (through indiscriminate bombing, sequester of oil assets, or treatment of prisoners along the lines established at Guantanamo Bay) would be followed by a reference to the ICC prosecutor. A legal 'shot across the bows'.

Mark Thomas' programme may not have altered government thinking or military planning, but the warning letter put the Ministry of Defence on notice. It was enough to frighten members of the general staff too. The most senior officers in charge of the invasion ordered that lawyers, members of the legal corps, be at the front line. They would make sure the laws of war were respected. It was the troops' insurance policy.

The letter also placed Shiner at the forefront of legal responses to the war.

8

WHEN THE INVASION OF IRAQ began in March 2003 and then ended soon after with dramatic haste, Phil Shiner began to hear rumours from Iraq about possible unlawful killings by British troops. The scandals were already surfacing from the mire of

Basra even before the Baha Mousa incident. Stories about ill-treatment at British bases appeared from some of the news agencies. Most prominent was the Camp Breadbasket affair, which came to light in peculiar circumstances.

On 28 May 2003 a soldier from the Royal Regiment of Fusiliers took a roll of film to a photographic developing shop in Tamworth, Staffordshire. He had just returned from Iraq and, with a returning holidaymaker's eagerness, wanted to see his photos as soon as possible. The shop assistant who developed them that day was repelled by what she saw. She called the police. The photographs didn't depict desert scenes or the streets of Basra, memories of an exotic army excursion. They were portraits of naked Iraqis, some forced to pose as if having oral or anal sex with each other, others trussed up in netting with guns to their heads, yet more with soldiers posing as though about to club them. One Iraqi was shown tied to the forks of a fork-lift truck suspended several feet into the air, clearly terrified. Many of the pictures had grinning British troops in the background.

What the soldier who wanted these photos developed had been thinking is hard to figure. Did he imagine that they wouldn't be seen? Didn't he care? Did he believe they were innocent snaps, trophies which anyone would find funny?

The story broke on the BBC. The photos were published to uproar. Defence chiefs were 'appalled'. The soldiers responsible, bad apples who had lost control, the military said, would be court-martialled; the army would ensure that this behaviour was stamped out; there would be a swift investigation and prosecutions would follow, or so the media were promised by generals and politicians.

For some old hacks in Fleet Street, the photographs might have aggravated a distant memory. This wasn't the first time that British soldiers fighting in a grubby war against a demonised enemy had snapped trophy pictures. Back in the early 1950s, in the Malayan Emergency campaign, there had been a similar scandal. The *Daily Worker*, forerunner of the *Morning Star*, an integral part of the British Communist Party, published a sequence of pictures depicting a grinning British Royal Marine holding up the severed heads of communist guerrillas to the camera. There was a mild outcry then, and many claimed the photographs had to be fakes. An investigation found that they were genuine. The army command admitted it was standard procedure: they needed to ensure proper identification of all those killed in jungle skirmishes, it was claimed. Troops were instructed to bring back the heads of the dead if it was impracticable to return with the bodies. It was macabre, indefensible, but it was officially sanctioned. Few protested strongly, though. The matter barely registered in the House of Commons. A few questions were the extent of the uproar. The Secretary of State for the Colonies, Mr Oliver Lyttelton, confirmed that the practice of decapitation would be ceased. No disciplinary action was taken against the marine. He hadn't actually decapitated the guerrilla fighters (that had been done by an employed native tracker from Borneo) and there were no orders at the time banning decapitation. Did that exonerate all concerned? Apparently so.

Fifty years on, *The Times* ran the story of Camp Breadbasket not as an isolated incident but as another example of the British Army's collapsing reputation for professionalism and respect. It reported concerns expressed by Amnesty International about the treatment of prisoners. It repeated accusations that had been made that the British encouragement of the people of Basra to loot Saddam Hussein's Ba'ath Party buildings was contrary to the laws of war. It noted that there were investigations into the behaviour of Colonel Tim Collins (later disproved), that he had pistol-whipped an Iraqi prisoner and staged a mock execution. The litany of allegations made for a tableau of abuse and lack of regard for the Geneva Conventions.

Less than four months later, in the autumn of 2003, the first reports of Baha Mousa's death in custody appeared.

Amidst all these stories, Phil Shiner recognised that the warning letter delivered before the war even began was now being tested. If there was evidence that British troops were paying little attention to the laws of war, how many other incidents of abuse were there, incidents which never made it to the newspapers? And where in the government and army responses was there an acknowledgement that maybe the whole system was flawed, that the preparations for war and the administration of the occupation ignored the humanity of the people of Iraq? Perhaps the 'bad apple' theory was simply not credible. In which case, how would the system be brought to account?

Shiner wanted to travel to Iraq in early 2004 to see for himself whether there were cases which were not being represented. But he was prevented from obtaining a visa. Iraq was too dangerous. Instead he instructed an

Iraqi citizen to whom he had been introduced to travel to Iraq and act as his representative. If anyone had a complaint about their treatment at the hands of British forces then he could take their instructions on Shiner's behalf. It wasn't a matter of seeking compensation. Shiner wasn't involved in that type of litigation. That wasn't his speciality. He was a public law expert, bringing judicial review proceedings against public authorities in order to change their practices, to highlight wrongdoing. This was standard in public law: actions brought to challenge decisions on a matter of principle. Sometimes, it was referred to as 'public interest litigation'. The individual brings the action not only to have their particular case reviewed by a court but also to set a precedent for others. A favourable judgment can alter the whole legal landscape, changing the behaviour and policies of local and central government and all those authorities purportedly operating for the benefit of all.

The Baha Mousa case would have been no different as far as Shiner was concerned. The clear injustice of Mousa's killing and the ill-treatment he and the other detainees received deserved a full and proper investigation.

Wasn't the RMP already doing this? Not according to Daoud Mousa.

9

I N January 2004 Daoud Mousa was waiting for news about the investigation into his son's death. He began to think there was something wrong. Were the authorities really taking the case seriously? His doubts were confirmed when he tried to encourage Sergeant Gillingham, the RMP liaison officer, to come and take a statement from him. After all, ever since Daoud Mousa had been informed about the death of his son, he'd let it be known that he had evidence to offer. He'd been at the hotel at the time of his son's arrest. He'd witnessed the theft of money by one of the soldiers. He'd reported it to Lieutenant 'Mike'. He suspected that it was this intrusion that had prompted the soldiers to single out his son for particularly rough treatment. He believed that what he'd seen on that day was relevant for the inquiry. He was an eyewitness, even though only to a small part of the story.

When Daoud was called to Basra airport on 5 January 2004 to receive an update on the investigation, he mentioned again what he'd seen at the hotel. Shouldn't his testimony be recorded properly? he asked. Neither Sergeant Gillingham nor the SIB officer present knew anything about this. They promised to contact the case investigator in the UK, WO Spence, and find out.

It took Sergeant Gillingham several days to track down Spence. When he asked about Daoud Mousa's evidence, Gillingham received a strange response. Spence reportedly said Mr Mousa couldn't have been at the hotel when his son had been arrested. There had been a full cordon that morning and everyone *inside* had been detained, Spence said. If Daoud Mousa had really been there he would have been taken into custody with the others. Spence didn't claim Mousa was lying, but it amounted to the same thing.

And then Spence was recorded as saying that in any event the suspects in the case hadn't been involved in the raid on the hotel. This would have been an extraordinary claim. It meant that the SIB weren't interested in any member of Anzio Company, who, of course, had carried out the arrests. But that couldn't have been right. Although Corporal Payne was the chief suspect, Spence knew quite well that a number of Anzio Company were present when Baha Mousa died. They were the guarding unit and had been since the evening of 14 September. In fact, Spence had seen Private Peter Bentham arrested on suspicion of murder, a suspicion yet to be formally refuted, and had interviewed several other men who had witnessed A Company soldiers help Payne restrain Baha Mousa in his final moments: privates Cooper, Reader and MacKenzie, and LCpl Redfearn. These men had all been at the hotel as well. It was absurd to say the arresting soldiers had nothing to do with the death. Either Sergeant Gillingham had misinterpreted what he had been told over the telephone, or WO Spence had fundamentally misunderstood the case he was investigating.

On 13 January Gillingham spoke to the SIB officer now in command in Basra and relayed what Spence had said. Gillingham was instructed to report to Daoud Mousa, to say that the details of the theft had already been included in the case file and the matter dealt with. This was strange too. If the matter of the theft was accepted, then surely Daoud Mousa had to be telling the truth about reporting it to Lieutenant 'Mike', whoever he might have been? And if that was accepted then he had to have been in the hotel, something which Spence had reportedly said wasn't possible.

Sergeant Gillingham spoke with Daoud Mousa the next day. He told him not to worry. The theft was being addressed, but no statement would be needed from him at that stage.

Daoud questioned the reasoning. Surely the motive for the ill-treatment of his son was relevant? Gillingham understood what Mr Mousa was claiming but as there was no corroborating evidence supporting the theory it was no more than that: a theory. Daoud said he intended to speak to the press about the case. And that was the end of the liaison.

It was hardly surprising that Daoud Mousa felt the investigation was neither full nor proper. When he was put in touch with Phil Shiner's representative, Daoud was ready to explore any possibility to obtain some kind of effective response from the British. Something had to be done. The only avenue left open was a legal challenge in the UK courts. It would be pointless to use the Iraqi legal system. British troops were operating under a general immunity from local prosecution. And who would listen to a judgment from a hearing in Basra anyway? Phil Shiner and his firm offered the only hope.

10

THE *DAILY MAIL*'s ATTACK ON the 'day of shame' and the part played by Phil Shiner in early May 2004 provoked more than just press criticism. A death threat arrived in Shiner's post the day after the headlines. Abusive emails followed. It was the start of a campaign of menace and intimidation that would continue for many months. Letters signed and unsigned arrived. Pathetic lines of outrage and vitriol written by authors barely literate, or so the messages suggested. Shiner's allegation of murder and the lodging of legal papers seemed to have inflamed individuals to react with spite. He was seen as a traitor, although what form the treachery took was far from clear. One anonymous letter said 'You are the epitome of everything wrong in this society . . . I really do hope that someone up in B'ham gives you a fucking good kicking you deserve, and stamp on your intellectual head. You are nothing but life's detritus. CUNT!' An email simply read 'COCKSUCKER!!!' Another letter warned 'We will firebomb your home and offices.'

Despite the threats, the action on behalf of Daoud Mousa and others continued. The process was inexorable, provided one knew the system. And that was Shiner's specialism. His twenty-odd years in the law had taught him the value of understanding procedure and playing within its margins.

On 13 May 2004 Shiner and the Ministry of Defence were called to the High Court. Permission was granted for a hearing on the issues. A date in July that year was set down. Arguments would be heard as to why UK law should apply to the killing of Baha Mousa and the other cases brought. The issue was straightforward, at least on the surface: did the British forces exercise sufficient control over the territory of south-east Iraq for the Human Rights Act to apply?

Although the issue may have been straightforward, the answer wasn't. There was precedent to suggest that the British courts had no right to look into these matters. They had taken place outside their jurisdiction and therefore it was for other tribunals to consider. The counter-argument also had its support; if it could be proven that the UK had effective control of an area as if jurisdiction had been assumed, then UK law should apply. Of all the cases lodged by Shiner the Baha Mousa affair was the strongest in this category. As Baha had died *within* a British base the test of effective control seemed easy to meet. Although the other cases also involved the killing of Iraqi civilians, this geographical distinction made the Mousa case special.

The flurry of press interest in the Baha Mousa investigation in May 2004 and all the reports seeping into the public domain of ill-treatment and then the judicial review application lodged by Phil Shiner's practice should perhaps have concentrated minds at the SIB. Their investigations were now just as much under scrutiny as the soldiers who inflicted harm on the detainees. If they weren't careful their reputations could suffer. The army could have its name sullied even further.

WO Spence and his team came under increasing pressure to complete their investigations. The legal division asked for all the witness statements to be sent to them for review. They had come back demanding more information. It meant re-interviewing most of the men who had already been seen once, twice, or even three times.

But things weren't as simple as when they were in Iraq. The soldiers had scattered. It was like trying to corral a field of rabbits. They would never stay still. Some had left the army, others had changed unit. Contact numbers didn't work. Private Lee was on sick leave. Aaron Cooper wouldn't answer his mobile phone.

It was difficult enough trying to see these people again, but others from Anzio Company hadn't even been spoken to yet. Now that members of the Hollender multiple were implicated (Fallon and Crowcroft were suspects after Private Lee's information) the numbers to be interviewed had expanded even further. More SIB officers had to be assigned to the case just to fulfil the legal division's inquiries. Were they enough? How much was being missed? Were they able to cope with the complexity of the case?

And then, as if in response to unspoken prayers, the SIB were handed a remarkable piece of evidence. It was a short video clip. It was dated 14 September 2003. And it was taken *inside* the detention facility of Battle Group Main.

It was a member of another company of 1QLR who brought the video to the SIB's attention. Corporal David Schofield had been in Basra with the rest of the battalion. He was a section commander, a long-serving member of the TA, and part of the community liaison group, the one which had been headed by Captain Dai Jones before he was killed. Schofield had visited Battle Group Main on 15 September 2003 and had seen the detainees there. He had used the Portaloos outside and had looked in through the windows just as so many had done before him. He had seen punching and kicking. When he'd returned to his vehicle he'd reported what he had seen to his commanding officer, Captain Good. The captain had gone over to the detention facility but returned soon after saying everything was quiet now. They had left the base and that was the end of the matter as far as Schofield was concerned.

By the time the SIB finally came around asking questions, in the autumn of 2004, they were merely trawling through all those who may have been at the base at the time of Baha Mousa's death. But apart from his own observations he had something else to give the investigators, he said. He had recently been asked to act as recruit trainer and wanted to show potential recruits what operational tours were like. He was going to compile a PowerPoint presentation. Thinking back to his days in Iraq, he thought he would send a message to fellow members of the battalion, those who had served in Iraq in 2003, to see if they could hand over some good material. He was sent a disk and on the disk, nestled amidst the usual photographs of dusty streets and checkpoints and laughing troops, there was a short video clip about a minute in length. Schofield opened it and must have been transported to that hot September day, the previous year, when he'd looked inside the detention centre at Battle Group Main.

The film depicted the detainees he'd seen, backs to the walls of one of the centre's rooms, their heads covered in sandbags and a thickset soldier screaming and swearing at the prisoners, pulling them to their feet when they dropped to the floor. The soldier in the clip was Corporal Donald Payne.

Schofield handed the disk to the SIB. It was like looking into the belly of a crime.

11

THE CIVIL CASE PURSUED BY Phil Shiner and his firm provoked considerable anxiety in Whitehall in July 2004. Baha Mousa's death posed the most significant problem for both the government and the army. There could be little doubt that Mousa and the other detainees had suffered abuse. That couldn't be denied, not seriously. Men had been arrested on suspicion of murder. And military detectives and lawyers were expressing concerns about the whole interrogation process. At the same time, though, there was absolutely no desire to submit to legal scrutiny every action of the armed forces in Iraq, or anywhere else for that matter. It could open up a secret world of intelligence and security that no one in government wanted exposed. If the call for a public judicial inquiry, which was at the heart of Shiner's case, was allowed, the delicate mesh of counter-terrorism and inter-agency cooperation, particularly with the USA, could unravel. There was no choice but to defend the civil action and deny the right of individuals in Iraq to look behind the screen of military practice.

It was a contradictory position for the SIB investigators. And it must have placed increased pressures on them to sort out the Mousa case as soon as they could. Any delays would only play into the lawyers' hands. It would suggest that proper inquiries hadn't been adopted. It would suggest that the army had no interest in finding out the truth and who was responsible.

Before two High Court judges in July 2004, Shiner's team, including leading barristers Rabinder Singh QC and Mike Fordham QC, presented the arguments for legal scrutiny of Baha Mousa's death. The call was for an 'effective official investigation' to be completed. State authorities were obliged to take reasonable steps to secure evidence in order to identify the perpetrators of any killing. Investigators had to have practical independence from those implicated in the alleged offence. It was submitted that the SIB investigation into Baha Mousa's death was suspect on both counts. The scene-of-crime examination hadn't been completed as soon as possible; witnesses hadn't been seen immediately; and the Royal Military Police officers were too bound up in the whole army structure to be truly independent of those they had to investigate.

The Ministry of Defence lawyers (including Professor Christopher Greenwood QC who was one of the few British academics to write that the Iraq War in 2003 was not in contravention of international law) countered these arguments. They suggested that the European Convention on Human Rights, upon which the standards of investigation were based, couldn't apply in Iraq. But if the Convention did apply, they argued, then its provisions *had* been respected. They submitted statements by Brigadier Moore, the commander of the brigade within which 1QLR operated, and Captain Nugent, head of the SIB in charge of investigations for the month or so after Baha's death in custody. Brigadier Moore explained the way in which all 'incidents', killings of Iraqis, would be assessed and investigated if the army's Rules of Engagement had been breached. Captain Nugent swore that the Baha Mousa case had been properly investigated. Inquiries had been concluded and a report submitted to the chain of command by early April 2004, her statement said. The investigation had been as effective as possible in the circumstances, she claimed.

This wasn't strictly correct. It was true that the case file had been sent by WO Spence to his superiors in April, but investigations had certainly not been completed. No charges had been levelled yet. Beyond Corporal Payne, they had no clear picture who would be prosecuted. Many men had been accused of assaulting the detainees at some point but who could be accused of killing Baha Mousa?

The High Court judges deliberated at length, finally reaching a judgment in December 2004. Although they concluded that for all sorts of good practical and legal reasons the Convention couldn't apply generally to the armed forces operations in Iraq, an exception had to be made for someone who had been held in a British military prison. The authorities couldn't claim that they didn't exercise control over such installations. The European Convention standards must therefore apply.

Regarding the Baha Mousa case the judges ruled that it was 'difficult to say that the investigation which has occurred has been timely, open or effective'.* They noted that the family hadn't been properly involved in the process. They criticised the delay in naming and prosecuting anyone for the killing. It was a damning conclusion and a major shift in legal thinking. By challenging the adequacy of the whole military investigation it forced the army to think more carefully about the abuse of the detainees in 1QLR's care. It must have sent a few shivers of concern through army command.

The Army Prosecuting Authority at the Ministry of Defence quickly assembled a legal team, including experts from the Bar led by Julian Bevan QC. They decided that the evidence against MacKenzie, Redfearn, Cooper and Rodgers was too flimsy to warrant public prosecution. They would concentrate on more likely culprits. Four potential defendants were identified, with Corporal Payne the key suspect. He was in the frame whichever way the case was examined. The Schofield video displayed him in the very act of ill-treatment so a charge of inhuman treatment would have seemed relatively straightforward. But could they prove murder? Various witnesses had spoken of Payne's struggle with Baha Mousa and some said they saw kicks and punches delivered moments before the collapse. Private Cooper had also said Payne hit the prisoner's head against a wall. Several could testify that Payne had his knee in Baha Mousa's back and may have been tugging on his neck. With the post-mortem pointing to positional asphyxia as a cause of death, the victim's ability to withstand the sustained assault degraded by the prolonged ill-treatment he'd received, they must have thought some charge of homicide could be sustained.

Murder, though, was a difficult proposition. The lawyers knew they would have to prove that Corporal Payne's actions were unlawful and he'd intended to kill or to cause grievous bodily harm. That was the test. If Payne could convince a court-martial panel that there were some mitigating circumstances or his intention was only to inflict 'serious' rather than *really* serious bodily harm, then manslaughter would be the more appropriate charge. So, what was it to be?

They must have deliberated over that for some time. It would have been galling to read those accounts of the determined assault upon Baha Mousa, the kicking and the evidence that Payne had bashed Baha's head against a wall or floor, and *not* charge him with murder. No one would believe that really serious harm wasn't intended. Then again, any court-martial jury

* The case was called Regina (Al-Skeini and others) v. Secretary of State for Defence.

would consist of serving army officers. They would hear that Payne thought Baha Mousa was trying to escape, that he used force to stop him. If that was believed, then maybe the panel wouldn't be convinced that Payne meant to inflict really serious injuries. Excessive force in his restraint, yes. But that might not be considered enough to amount to murder in their eyes. Manslaughter would be safer. Even so, if Payne could show that his actions were lawful, that he was justified in using force to prevent escape, then a defence against manslaughter could succeed as well.

This was where the prosecution relied on the medical evidence. It meant that Baha's death occurred after a series of assaults, culminating in the restraint Payne had used. If the prosecution could convince the panel that this was right, then whether or not Baha had been trying to escape by removing his hoods and cuffs, Payne's actions taken as a whole were unlawful and dangerous.

The prosecution team agreed to proceed on this basis. It meant that they would have to prove that Corporal Payne was responsible for the continual beating of Baha Mousa and his final restraint.

Would this be enough? The case wasn't just about Baha Mousa, after all. The other detainees had been subjected to equally intense abuse. If Payne was the only accused then he would look like the scapegoat, the patsy, the sacrifice. It wasn't difficult to imagine what he could make of that in court and what the media would say as well.

A tactical decision was taken during the early days of 2005 to pursue only the strongest cases. The prosecution team must have believed that they didn't have enough to go after privates Slicker, Cooper and Bentham. Privates Crowcroft and Fallon and Corporal Stacey were deemed better targets. There was the testimony of Private Lee. And it was known that these soldiers had been on their own with the detainees for a substantial part of the first day of imprisonment. This was when the conditioning of the detainees had been introduced. They must have spent several hours enforcing the stress positions. For the prosecution this was enough to set them apart from the men of A Company.

Of the officers, Lieutenant Rodgers was perhaps lucky to be left out of it, but then most of his multiple had given statements saying he wasn't at the facility much and that in any event none of them had assaulted the detainees. They had just followed instructions to carry on with the conditioning. That wall of evidence would be difficult to break down.

Major Peebles? It was clear that he had presided over the whole conditioning cycle with SSgt Davies, the tactical questioner. Those two must have known what they were doing was wrong. The hoods, the stress positions, the sleep deprivation couldn't reasonably be justified. Whether they knew of the punches and kicks or not, they had introduced and maintained the system of abuse well beyond the time limits when they were allowed to keep prisoners at Battle Group Main. The charge against them would be neglect of duty. But why wasn't the other tactical questioner, Sergeant Smulski, included? Perhaps the fact that he'd arrived much later in the process and took his cue explicitly from SSgt Davies was enough to absolve him. He could claim that he 'was only following orders'.

Payne, Fallon, Crowcroft, Stacey, Peebles and Davies: would such a cast of defendants be sufficient? Would it satisfy everyone shouting for something to be done? Who, after all, was responsible for everything that happened at Battle Group Main? Shouldn't Colonel Jorge Mendonça (as he had become following promotion in 2004) bear some of the blame? Shouldn't he take ultimate responsibility for the behaviour of his unit? And hadn't he been personally implicated anyway, if only by failing to check that the system of detainee handling was appropriate?

The prosecuting team examined this very carefully. It was no small thing to recommend charging a senior officer of the British Army in relation to the commission of war crimes within his battalion. Would they have been leaned on by the Ministry of Defence or by the Attorney General? Would there have been political intervention to encourage such a prosecution? It might have helped send a message that the government wasn't afraid to root out bad apples in the forces no matter how senior. There would be no whitewash in Whitehall, or something similar may have been the underlying sentiment. Even commanders could be held accountable, it would have implied.

12

THE PRESSURE TO GO AFTER a senior officer became acute in February 2005. It was then that the Camp Breadbasket trial took place. Those soldiers who had been photographed with an Iraqi tied to a fork-lift truck and who had forced prisoners into simulated sex acts appeared before a court martial in a British base in Osnabrück, Germany. Lance Corporal Mark Cooley, Corporal Daniel Kenyon and

Lance Corporal Darren Larkin were accused of a host of crimes. Cooley was charged with 'disgraceful conduct of a cruel kind' for being pictured driving the fork-lift truck with an Iraqi tied to the front prongs. He was also accused of simulating a punch seen in another photograph. Kenyon, the most senior soldier of the three, was accused of failing to report what had happened and aiding and abetting others in assaulting the prisoners. Larkin had been pictured standing on the prone body of one of the Iraqis held at the camp, an action which incurred a charge of assault. A fourth defendant, Gary Bartlam, the soldier who had taken the trophy photographs, had struck a deal. He had pleaded guilty to a number of charges and was dishonourably discharged from the army as well as sentenced to eighteen months in a youth detention centre before the other defendants even reached court.

The photographs couldn't have been more damning. It was impossible to give an innocent explanation for them, although LCpl Cooley tried. He'd argued that he was only moving the prisoner out of the sun. He wasn't believed. But the soldiers' general defence held significantly more weight. All four argued that they were given orders to act like this. It was claimed in court that a Major Dan Taylor, the officer in charge of the camp, had become fed up with looters entering the base and stealing supplies. The accused alleged that they had been given orders by Major Taylor that anyone captured inside the base should be 'worked hard' to deter them from coming back. It was part of Operation Ali Baba, as it was called. They took their instructions at face value, using their imagination to decide what would be appropriate. No doubt Major Taylor would have challenged their interpretation, but even the order he allegedly gave would have been against the rules of war. Prisoners shouldn't be 'worked hard' as some form of retribution. Despite this, Major Taylor wasn't standing accused. Why not? 'Shit rolling downhill' the squaddies of A Company would have said.

The claim was given some credence when General Sir Michael Jackson, head of the army, announced after the trial that Major Taylor might be subject to some kind of internal disciplinary process. However, the key charge levelled at the Army Prosecuting Authority that the junior soldiers brought to court martial had been made scapegoats with more senior officers escaping criminal sanction remained credible. Rumours began to surface that government officials weren't too happy at such selective prosecution and were unimpressed by the apparent lack of rigour in the army's prosecution of its own. It all looked a little too cosy. Kicking a few young squaddies out of the forces had the appearance of tokenism.

General Jackson appeared sensitive to the critique. His apologetic statement released after the Camp Breadbasket affair had mentioned that he was appointing a senior experienced officer to track other prosecutions and make recommendations for action. It was an acknowledgement of failure, but it didn't go as far as it could have done. The case against the Camp Breadbasket soldiers had suffered from a lack of evidence. None of those victims pictured in the photographs was brought to testify. The APA said they couldn't be found. But three of the men had been in contact with Phil Shiner's firm and given initial statements about their treatment. Shiner passed the information on to the Attorney General and the APA. He was told to leave it at that and say nothing. Meanwhile, the day after the trial ended, the *Independent on Sunday* published a story that despite the RMP's assertions that they couldn't locate the victims of Camp Breadbasket, the paper had found them with ease. And the victims' accounts of what had happened to them were far worse than even those related at the court martial in Osnabrück. As in the Baha Mousa case, there were allegations of beatings and humiliation. They seemed to confirm the 'work them hard' order had been applied with enthusiasm.

It all pointed to a deeper culture of abuse than General Jackson would admit and an army police force which failed at the most basic level of evidence collection. The Attorney General was reported as looking for reform in the way the army prosecuted crimes. Whether he also called for a change in approach to be reflected in the ongoing Baha Mousa case is unclear. Was it coincidence that Colonel Mendonça now became a target for the APA?

Brigadier Robert Aitken, the officer appointed by General Jackson after the Camp Breadbasket trial to look into the conduct of prosecutions, reportedly wrote to battalion commanders who had served in Iraq. According to the *Daily Telegraph*, his letter said the 'military discipline system' was receiving 'a bit of a hammering'. It asked 'whether our officers behaved to the highest standards'. And it requested 'evidence of officer behaviour in Iraq which I could use.' It wasn't clear what this *use* meant, but the underlying intent was sharp enough.

Colonel Mendonça was one of those sent the letter. He replied positively about those who had been under his command. He stressed the uniquely dangerous, even harrowing, circumstances of the battalion's time in Basra. The immense restraint his men showed and the discipline they maintained did credit to them all. And he added his own political spice: he told the brigadier that his officers were hampered by the need to make good the 'Coalition's complete failure to plan for the aftermath of the war'. He wasn't about to criticise any of his men in these circumstances. He didn't believe he had good cause to do so.

Whether there was any political pressure that followed this correspond- ence or not, the APA team began to think how they could connect Mendonça with some identifiable crime. In the absence of any direct evidence placing him in the detention facility whilst the detainees were suffering abuse, they had to consider neglect of duty. But what was that duty and how had he been negligent?

The legal experts set about building a case on a simple assumption: the commanding officer of an army unit should always take responsibility for everything that happened under his command. If one took that principle to extremes, then no matter what crime had been committed, almost regardless of context and circumstance, a battalion commander would have to answer for the conduct of his men. This was stretching plausibility too far. In Mendonça's case, though, the fact that he was sleeping and working within a hundred metres of the detention facility where Baha Mousa died, where all that screaming and shouting occurred, where detainees were taken across the compound for interrogation over a period of forty-eight hours, placed him right on top of the scene. He was close enough to touch it, to smell it, to hear it. It brought culpability that much closer to Mendonça's door.

In June 2005, as the festering political tension induced by all those stories of army misconduct continued, the SIB sought the opinion of a specialist. Brigadier Robert Scott-Bowden, director of infantry, was asked to sum up the responsibilities of a battalion CO. He said they incorporated the 'training and safety, security, discipline and education, welfare, morale and general efficiency of the troops under his command'. Although he said he would expect the CO to delegate prisoner handling, *all* responsi- bility could not be avoided. He had to ensure those given the job of running a detention facility, for instance, were well aware of their legal duties. That included observing the laws of war. The brigadier said the

CO must put in place 'an organization that is adequately trained and sufficiently resourced'. These obligations couldn't be shifted wholly to junior officers.

This was enough. The APA had sufficient cause to pursue Colonel Mendonça, or so they believed. A charge was drawn up: negligent performance of his duty by failing to take such steps as were reasonable in all the circumstances to ensure that Iraqi civilians being held at the temporary holding centre under his command were not ill-treated.

13

ON 19 JULY 2005 LORD Goldsmith, the Attorney General, made a public announcement that British soldiers would stand trial for war crimes. Corporal Donald Payne (thirty-four years old) was accused of manslaughter. He, Private (now Lance Corporal) Wayne Crowcroft (twenty-one) and Private Darren Fallon (twenty-two) were charged with inhuman treatment. Corporal (now Sergeant) Kelvin Stacey (twenty-eight) was alleged to have assaulted one of the detainees. SSgt (now Warrant Officer) Mark Davies (thirty-six and a member of the Intelligence Corps), Major Michael Peebles (thirty-four) and Colonel Jorge Mendonça were accused of negligently performing their duties.

As soon as the charges against the 1QLR men became public a certain language infiltrated the copy of journalists: 'witch-hunt', 'scapegoat', 'politically motivated'. There were statements from Conservative MPs denouncing the prosecutions. Ex-soldiers voiced their disquiet. And in an unusual open expression of discontent, the then commanding officer of the Queen's Lancashire Regiment, Brigadier Geoffrey Sheldon, issued a statement in support of 1QLR's former commander and indeed the whole regiment. It was an unprecedented step.

He wrote 'It was Colonel Jorge Mendonça, then the commanding officer, who, as soon as he learnt of Mr Mousa's death, initiated the formal inquiry that has now resulted in these charges being brought. It is, therefore, particularly difficult for us to learn that Colonel Mendonça must himself now answer charges as a result.'

Brigadier Sheldon's extraordinary statement was a defence of the collective integrity of his regiment, and by implication, the whole British Army. Colonel Mendonça was one of the best, he said. And the battalion he led was recognised as one of the finest to serve in Iraq in that chaotic time. If he could be denounced like this, with so much public fanfare and disregard for his reputation, which commanding officer could have any confidence in their position? Any allegation of abuse by one of the ranks could prompt a prosecution for neglect of duty, whether that duty had been properly delegated or not.

Some would have sensed this as a political moment. Not only did the army have to fulfil the demands of ideological hubris, it was now being asked to bear the responsibility for its consequences. Military intervention would never be without its brutal character. Even the most naïve politician had to know that. If the officer corps was to be criminally liable for every manifestation of that inevitability, who would dare to assume command in the future? And who would protect them?

The inclusion of Colonel Mendonça on the charge sheet transformed the shape of the SIB's case. Even though from the first moments of the investigation they had collected information about the chain of command and how that was supposed to function in the treatment of detainees, there was now a pressing need to define the obligations owed by each rank. It was no longer sufficient simply to establish the facts of abuse. The whole system of detention as operated by 1QLR under Colonel Mendonça's command had to be shown to be at fault.

A sequence of failures framed the argument. They centred on what Colonel Mendonça knew or should have known. As far as the lawyers were concerned, the neglect of duty arose because the CO *knew* conditioning was being undertaken. He knew this involved stress positions, deprivation of sleep, hooding and cuffing. And he knew that the detainees were being kept at Battle Group Main for much longer than orders decreed. Being aware of all this, he should have checked on their welfare, he should have checked how long the conditioning was being applied, he should have checked how they were being treated when the interrogations stopped. And when he found out about the death of one of the detainees, he should have been down at that detention facility making sure everyone inside was well. He didn't need to be implicated in the ill-treatment. That wasn't the point. His command status simply meant that as soon as he was told that there was a problem his priority should have been to ensure the remaining detainees' welfare. It wasn't enough to leave it to others, to assume someone else would attend to this.

These were the legal arguments. But the political dimension refused to be suppressed. Stories criticising the damning of Mendonça appeared. Mendonça's wife talked to the press. She was appalled at how one moment her husband could be awarded the DSO for his work in Basra and the next prosecuted for failing in his duties. Her life was being torn apart under the strain of watching her husband shamed when she was convinced no such shame was justified.

She wasn't alone in her feelings. Family of soldiers accused in unrelated crimes in Iraq believed they were the victims of political interference as well. When the Attorney General had announced Colonel Mendonça's prosecution he had also publicised another case. Four members of the Scots and Irish Guards were accused of killing Ahmed Jabber Kareem, a fifteen-year-old boy, in May 2003. Kareem had allegedly been arrested on suspicion of looting and had been thrown into the Shatt-al-Arab waterway to teach him a lesson. The boy was asthmatic and couldn't swim. He disappeared under the water and didn't resurface. His body was discovered two days later. Others thrown in with him also spoke of being beaten before they too were pushed into the canal. The soldiers were to be prosecuted for manslaughter and again no officers were charged.

The court martial of Sergeant Carle Selman, Lance Corporal James Cooke, Guardsman Joseph McCleary and Guardsman Martin McGing took place before the trial of the 1QLR case was heard. In the early summer of 2006 the prosecution claimed that there had been a common plan to punish looters. It had been called 'wetting'. These four soldiers executed the plan, or so it was alleged. The case lasted over a month. All were found not guilty.

The accused spoke freely after the verdict. Like Mrs Mendonça their vocabulary was sharpened by the word 'betrayal'. As with the Camp Breadbasket case, they pointed to orders they had received that led to Kareem's death. McGing was quoted as saying 'The idea was that once they were wet they had to walk home to change their clothes. It stopped them looting. We were doing what we were told to do and being led to do.' Guardsman McCleary told the *Daily Telegraph* 'Looters were every-where, there were too many of them for us and it was difficult to know how to handle them. We were told to put them in the canal. I was the lowest rank. We were always told we weren't paid to think. We just followed orders. We had a job to do.' He said, with ironic use of cliché, the army had 'hung him out to dry'. It was yet another case of shit rolling downhill.

The death of Kareem went unpunished and all involved were left tarnished and bitter.

14

MANY MONTHS BEFORE THE COURTS martial for the killing of Ahmed Kareem and Baha Mousa in 2006, the Army Prosecuting Authority and the whole system of military criminal justice experienced another catastrophic setback. After the dissatisfaction evident from the Camp Breadbasket prosecutions, the APA needed an uncritical success.

Seven members of the 3rd Battalion of the Parachute Regiment appeared before Colchester Barracks Court Martial Centre in late 2005 accused of murder and violent disorder. The prosecution claimed that they had killed an eighteen-year-old boy called Nadhem Abdullah in May 2003, shortly after the war had been declared won. Whilst on patrol in a village near the Iranian border, they had chased a car believed to be involved in smuggling and then stopped another, a taxi, when the other car had sped off. Nadhem Abdullah was in the front passenger seat. Everyone in the taxi was told to get out and lie on the ground. The soldiers called them 'Ali Babas', which seems to have been a semi-official army term for thieves. And the paras then laid into the men from the taxi, hitting them with rifle butts and fists and boots. The soldiers left them all injured on the ground. Nadhem Abdullah was later taken to a doctor, the prosecution said, but there was no neurosurgeon available. The family arranged a car to take the eighteen-year-old to one of the Basra hospitals. Nadhem died on the way.

The evidence offered by the prosecution was patchy. No Iraqi could identify any of the accused paratroopers. The soldiers had been questioned and they had denied that anything had happened. There had been no post-mortem. But blood had been found on the butt of one of the soldiers' rifles which matched the DNA of a family member of the dead boy. And records suggested that no other British patrol had been near the vicinity at the time of the incident. It was a flimsy case but the prosecution went ahead nonetheless.

In November, after several weeks of the hearing, the judge advocate called a halt to the trial. The evidence was simply insufficient. Worse, it was deeply tainted. Several Iraqi witnesses admitted that they had only agreed to testify when they were told they would be paid $100 a day for their time. Others admitted that they had lied, that they had been encouraged to lie by the dead boy's family, or that they had exaggerated what had happened to demand compensation. It was a sordid disaster for the prosecution.

After the Camp Breadbasket farrago and now these unsuccessful prosecutions, it was a wonder that the Army Prosecuting Authority felt comfortable in proceeding against those accused in the Baha Mousa case. Except the evidence was a little richer. They had a body, a post-mortem, photographs of livid injuries, witness testimony from serving soldiers, clear accounts of the treatment, a video of the abuse in action, a system of 'conditioning' that purportedly contravened firmly established army directives, admissions of breaches of orders and well-known war conventions, and accounts from victims which were plausible and consistent.

The case was different. It had to be pursued. The lawyers would simply have to ensure that it was presented with care. If everything went according to the evidence they had in their files, then a conviction was not only possible but highly likely.

But . . .

As any good lawyer will say, a case is only as good as the evidence the court can see or hear or touch. It doesn't matter how morally outrageous a crime may be. Nor does it matter how obviously guilty a person may seem. When it comes to securing a conviction, all that counts is the quality of proof presented. That is supposed to be the ascendant virtue in British courts. Any good lawyer will also say that proof consisting of eyewitness testimony provokes both opportunity and threat. A capable witness can communicate not only facts but also emotion. Juries and judges can have their understanding of a case transformed by a visceral account of a violation, one they can believe and feel. Their sympathies can be seized so that all other evidence is viewed with a mind already primed.

And the threat? Eyewitnesses also provide a chance for defence lawyers. If a witness appears erratic, flaky, uncertain, prone to exaggeration, then the jury's sympathies can be displanted. These are the moments lawyers love to incite. With studied skill rather than malice, 'breaking' a witness appeals to the intellectual conceit of the advocate. It can be a triumph, a vindication of one's acquired art. The person subjected to the process, however innocent or however devious, becomes a puzzle to crack. Watching him or her fracture under cross-examination can be painful. If the advocate gives any thought for their well-being then the art fails. It is the client who matters.

Opportunity and threat. By the summer of 2006, after countless re-interviews of dozens of soldiers, after months of careful cataloguing of documents, after case meetings and drafting meetings, after analysis and preparation engaging a multiplicity of legal minds, the case was ready. The court martial could begin.

Part 3

THE COURT MARTIAL

BATTLE GROUP MAIN WASN'T A *death camp or a torture centre. It was a military base where idle cruelty became possible. There was no systematic plan to kill or eradicate an enemy. That doesn't mean war crimes were not perpetrated within its walls. Accusation and judgment remain appropriate. If those international standards, the Geneva Conventions and their like, which are almost universally endorsed, are to be credible, any behaviour that breaches them must surely be scrutinised and those accused confronted. That is the fundamental proposition which shores up every war-crimes tribunal.*

The case against the seven British soldiers implicated in the death of Baha Mousa and the ill-treatment of the nine other detainees was no different. They faced the ubiquitous method by which their alleged crimes were to be judged: the trial. Whether or not this is a proper or effective or efficient method, it's the only one available. Some say the method is flawed. They say the trial is no more than a theatrical performance. It's a show, where posturing, play-acting and rhetorical manoeuvres flourish. The participants can be reduced to caricatures, labelled according to categories: victim, witness, accused, lawyer. All are called upon to conform to a pattern of behaviour controlled by custom, or written procedure, or simply expectation. What's said, how it's said, when it's said, are predisposed by rules followed with devotion and scrupulously policed. For many it's this quality that provides the trial with its legitimacy even if it suppresses the humanity of all those affected.

The long and fractured investigation into Baha Mousa's death ended with just such a trial. After three years of inquiry, the performers gathered in Bulford Military Camp on Salisbury Plain. The September warmth was beginning to dissipate, although it still felt almost balmy when the sun shone through the windows of the newly constructed court building. It wasn't quite the same as on the streets of Basra where the heat would be flattening creatures against shadows and the burden of body armour drew perspiration faster than one could drink. But it was comfortable, which was just as well as the military participants arrived in their straitening uniforms, buttoned to the neck.

There would be a multitude of characters appearing. Many of the surviving detainees, though not all, would come. Dozens of military personnel from the lowest rank to brigadier would assemble to give their accounts of now distant events. There would be army lawyers, trainers, intelligence officers, members of the Territorial Army who no longer served, staff officers talking about tactics and systems and protocols, medical doctors to interpret injuries and their likely causes, a padre too and a number of Royal Military Police officers to introduce evidence. The cast would run into hundreds by the time proceedings ended towards spring the following year.

And, of course, there were the seven men assigned the special role of 'the accused'. They would sit with their lawyers, remaining silent. Once they confirmed their names and pleas, they would not utter a word: no exclamation, no shout, no interjection. Theirs would be a role without a voice. Colonel Mendonça, the commanding officer of 1QLR; Major Peebles, the man in charge of interrogating the detainees; WO Davies, the tactical questioner; Corporal Payne of the regimental provost left to look after the detention block; Sergeant Stacey, LCpl Crowcroft and Private (now Kingsman) Fallon, members of the Hollender multiple of Anzio Company and the first to guard the prisoners.

The lawyers, some of the most senior barristers at work in the criminal law, would flap about these men like considerate carrion crows, worrying over their clients whilst maintaining more fraternal association with their own kind. They would chuckle amongst themselves, sharing a joke or a vignette of their travel to the court. Later they would stand and sit like erratic pistons within a chaotic engine, as only one would be allowed on his feet at a time. They would perform and speak for the soldiers sitting behind them, projecting their professional personalities into the chamber and towards the people who would stand in the witness box. Sharp-billed, masters of the conventions that guided them, so prac- tised in their art that they would perform with coordinated brilliance, they would throw acidic remarks at their fellow lawyers one moment and sit down with them for a chat about the cricket the next. This trial would be their show more than anyone else's. It was their arena and even the judge would bow to their collective skill. He might admonish them from time to time, but he probably knew most of them from old and he wouldn't constrain their performances. He recognised this was their territory. After the proceedings concluded and all those months passed, they would know each other better and might one day relive the passage of the case. They won't share any secrets or privileged information, but that wouldn't stop them talking in confidence. Their weeks spent in Bulford Court Martial Centre would be part of their professional bond.

Such is the character of British justice in practice. It wouldn't matter whether this was a case of handling stolen goods or fraud or brutal murder or systemic abuse of Iraqi civilians. The proceedings would still be pinioned to a protocol only the lawyers would understand in its entirety.

1

THE COURTROOM REEKED OF NEW carpet and wood polish. Lawyers and clerks and ushers wandered in and out, their footfalls and chat absorbed by the pine-panelled walls and blue-carpeted floors. The cushioning of noise was deliberate. Fearing that proceedings would be interrupted by low-flying military aircraft from nearby Boscombe Down, the builders had installed deep soundproofing, cocooning the courtroom in an echoless chamber. No outside clamour penetrated. No shouts of command to waiting defendants. When the doors were closed and proceedings began, the specially designed acoustics accentuated the merest sigh or quick breath or tapping of keys.

This was Bulford Court Martial Centre in September 2006.

The press were calling the case against the seven defendants a war-crimes trial. They were enjoying the circus. It was a unique occasion which would continue to fill headlines and bulletins for weeks. The very term 'war-crimes trial' evoked passions and memories. Editors could turn to those black and white grainy pictures of Nuremberg and Göring and Speer and all those Nazis and draw comparisons, however absurd. Readers would make the connection even if it wasn't pointed out to them.

For those taking part, the court martial had a more mundane feel. The advocates were ordinary criminal barristers. They would behave no differently than they did in their last appearance at the Old Bailey or Winchester Crown Court or Horseferry Magistrates or wherever else they practised. Julian Bevan QC was the lead performer. He commanded the prosecution, his Old Etonian manner charming those about him and offering a solid

personality in charge of the elephant of a case he had to present. Assisting
him were a number of lesser figures: Lieutenant Colonel Eble of the Army
Legal Service and two junior barristers. Between them they would
coordinate all the evidence and submissions the court would hear over the
succeeding six months.

And then there were the defendants' barristers. In truth they would
monopolise proceedings, queuing up on occasion to address a witness,
muscling each other out of the way or cooperating for greater effect, all
in the interests of their particular client. Like professional actors, they had
their reviews, comments on their past performances quoted with relish on
their chambers' websites. Even the most senior members of the Bar
appeared to need good publicity.

Tim Owen QC represented Corporal Donald Payne. He was a suave-
looking man, robust in build and fleshy-faced, but with a bitter wit. His
website quoted the assessment that he was 'as good as it gets' in public
law matters. His junior was Julian Knowles, although his acclaim was no
less meritorious, described as 'demonstrating all-round brilliance'. They
would share the cross-examination duties although it would be Owen who
addressed the crucial witnesses.

LCpl Crowcroft's barrister was Richard Ferguson QC. One of the
Bar's most experienced practitioners, Ferguson had acted in a string of
notorious criminal cases over decades. Normal retirement age had passed
several years previously, but that hadn't diluted his reputation for hypnotic
charm in the courtroom. He would die three years later at the age of
seventy-three and so this case was something of a swansong for him. He
was assisted by William England, a man who could claim wide experience of such cases given that he'd already appeared in the Camp
Breadbasket court martial. His expertise would be officially appreciated
some years later when he would be instructed by the Ministry of Defence
to represent them in the Baha Mousa Public Inquiry that would begin
in 2009.

Crowcroft's comrade, Kingsman Darren Fallon, was represented by
another doyen of the criminal Bar: Geoffrey Cox QC, Member of
Parliament for Torridge and West Devon, a Conservative, elected in 2005,
leading barrister and head of Thomas More Chambers in Lincoln's Inn,
London. His chambers' website noted he was 'highly regarded for his
versatility', and described his advocacy as 'flamboyant' and 'extremely

persuasive'. It was a wonder that he had the time to be in court. Mr Cox was a very experienced advocate, as his chambers' résumé would remind anyone who cared to look. He too needed an assistant: a Mr Cross was on hand.

Major Peebles was able to call upon another very senior barrister. Lord Thomas of Gresford QC was, and remains, another parliamentarian with a distinguished legal career. A member of the Liberal Democrat Party, he was appointed to the Lords in 1996, having failed in numerous general elections for the Wrexham constituency. Whilst the trial at Bulford proceeded he also acted as Shadow Attorney General and Shadow Lord Chancellor. Like Geoffrey Cox, he must have been an extremely busy man, dividing his time between Westminster and Salisbury Plain. Lord Thomas was assisted by a Mr Clark.

Then there were the advocates acting for Sergeant Kelvin Stacey: Jeremy Baker QC and Ms Fiona Edington, both of whom worked from the same set of chambers as Geoffrey Cox. Edington was the only female barrister taking a significant part in the proceedings. Given the male domination of the profession, mirrored in the military, this was no surprise.

WO Davies had one of the more prominent defence barristers available, Jeremy Carter-Manning QC, whose particular specialism was publicised as 'fraud' on his chambers' website.

And finally those representing Colonel Mendonça were Timothy Langdale QC and Bernard Thorogood. Langdale's plaudits on his chambers' promotional material were effusive. He was quoted, apparently without irony, as being 'more polished than Aladdin's lamp', 'incredibly adaptable', 'a marvel as an advocate', and 'like a champion boxer who has marked out every inch of the ring, he knows exactly where he is and precisely how to manoeuvre his opponent'. Thorogood's publicity was more subdued, but it recorded that he once served in the army on a short-service commission in the infantry. No doubt his experience would be useful.

Between them, this gathering of defence barristers could claim decades of experience in criminal proceedings. There could be no doubt that the smallest gap in the prosecution witnesses' evidence, the slightest contradiction or confusion, would be pierced by these seven QCs and their numerous able juniors. Collectively, if they *were* ravens, which they resembled in their black court gowns flapping down the corridors of the court, they would be called an 'unkindness'. That would be a quality many witnesses would experience by the time the case ended.

2

H OW MIGHT THOSE IRAQI DETAINEES flown from Basra to Britain have felt as, one by one and alone, they opened the door to the courtroom in that army base in the middle of the English countryside stepping Alice-like into that hushed chamber? As 'the dream-like child moving through a land of wonders wild and new' perhaps, passing the ranks of lawyers and clerks and soldiers, led into the witness box and sworn in with the help of an interpreter. Everything must have seemed eccentric, forbidding, deeply strange.

There is a strong tradition in English law that a fair trial requires particular conditions: an equality of arms, representation by someone skilled in the art of examination, the opportunity to test the evidence laid against a defendant. These are protections against 'unfairness'. But what 'fair' means in this context is particularly fluid. Fair to whom? Fair to the society where a crime has been committed? No. One might argue that all criminal cases represent 'society's' attempt to be treated fairly, but really this doesn't hold up to intense scrutiny. Fair to humanity? Hardly. When we speak of 'crimes against humanity' we might be referring to some human consciousness which is offended and needs to be addressed in a fair and reasonable way, but few would maintain this directs the conduct of criminal proceedings. Fair to the defendant? This is where the notion has some bite. In order to ensure that anyone accused of a crime is given a full opportunity to defend themselves and decisions are made according to evidence rather than prejudice (or should that be prejudgment?), safeguards are applied. This is the meaning of fair trial in most western legal systems. But fair to the victim? No. That may be a contemporary concern for politicians and newspaper columnists, but it isn't one for the courts.

Where does that leave those who have been victimised? Pretty much where the Iraqi men called to give evidence at Bulford Court Martial Centre were: 3,000 miles from home, in a foreign and unfamiliar environment, trooped before a bevy of advocates intent on applying a scalpel to every word uttered (and words refracted through interpreters), treated as suspicious and forgetful, unreliable and prone to exaggeration if not fabrication for their own monetary gain. If a fair trial is supposed to include them then there was little evidence of that in Bulford.

None of these matters concerned the prosecution in the slightest. A witness is a witness; nothing more, nothing less. An evidential commodity.

On 26 September 2006 Mr Bevan called the first detainee to give evidence for the prosecution, Mr Ahmad Taha Musa al-Matairi. There were few preliminaries before Bevan took the witness back in time, back to the hotel when the soldiers had arrived. Al-Matairi recalled the early-morning raid, the searching and then the pushing and shoving, as the hotel workers were forced on to the floor. But his testimony wasn't as clear as Bevan had hoped. Confusion began to penetrate the dialogue.

'Whilst you were lying on the floor did anything happen to you?' Bevan asked.

'He started to hit me,' al-Matairi said.

'Sorry?'

'He started to hit me.'

'*He* started to hit me?'

'He started to hit me and wouldn't allow me to raise my head and the officer was looking at him.'

'Who's "he"?'

'A soldier.'

'A soldier. Can you describe how the soldier hit you?'

'He would hit me with his boots.'

'And where?'

'On my head.'

'How heavy was the kick?'

'It was an insult kick.'

'An insult kick?'

'An insult kick.'

No definition was given for the expression 'insult kick', but the impression was of a tap, not a swinging boot. It somehow lessened the scale of abuse, gave it an almost innocent or trivial character.

Slightly perplexed but intent on focusing on more important matters, Bevan raised the issue of the weapons found in al-Matairi's hotel, the ones that brought out the change in atmosphere and sparked the viciousness in the soldiers, or so it had seemed. Bevan knew that this was a small fissure in the prosecution's case: if the defence could argue that the discovery of rifles and pistols and night sights entitled the soldiers of Anzio Company to become rough with the detainees as potentially dangerous terrorists then the sympathies of the panel could shift.

Al-Matairi confirmed that weapons had been found, but they were for self-defence, he said. It wasn't a convincing explanation. A rambling answer suggested sensitivity, wariness, vulnerability. The jury panel of army officers would no doubt have been asking themselves 'What if *I* found them in a search I was conducting? What would I think? And what would I do?' Bevan couldn't prevent these questions from settling in the minds of the panel. Nor could he erase the probable answers. All he could do was lay a more impressive image over any forming view. He had to shock with the abuse suffered, have the panel shout internally 'But I would never have reacted like that!'

Bevan reminded the witness of the moment when he and the other prisoners from the hotel were brought to Battle Group Main. Al-Matairi recounted his memories of the abuse that followed in a cascade of invective.

'They would come from behind and kick me on the knees . . . they were betting on me falling . . . he hit me on the kidney . . . struck on the back . . . beaten . . . continuously and repeatedly . . . I would ask for water . . . he would remove the sack and put hot water in my mouth . . . I would tell him . . . he would start to laugh . . . under the sack . . . I couldn't wipe away the sweat in my eyes . . . two sacks . . . I was about to suffocate . . . around twenty soldiers in the evening . . . they would hit us and laugh . . . they were celebrating beating us . . . it was like Christmas . . . nothing to eat on the first day . . . I was interrogated . . . they wanted to know where the one who escaped was . . . I didn't know . . . they gave me a minute to answer or I would be sent to Bucca prison . . . back . . . hit with a wood stick . . . they were playing karate, kung fu kicks . . . not allowed to sleep . . . I had a hernia . . . it swelled in the groin . . . it was the beating . . . I was hit there . . . I cried out in pain . . . the hernia became enlarged . . . I was seen by a doctor . . . at the base . . . I told him about the beating . . . he said the beating had to stop or I might die . . . I wasn't beaten but Baha was . . . he kept crying "Blood, blood" . . . his wife suffered from cancer and she had passed away six months earlier . . . he kept on crying "My children, my children, I am going to die" . . . we heard nothing more . . . the beatings stopped.'

The sudden injection of Baha Mousa's voice should have incited pity in the courtroom. But the emotional delivery by the witness, the ebullient movements of body and voice, facial expressions, even words which couldn't be properly translated, jarred with the British courtroom reserve. Pain was to be measured, quantified in this arena, not felt. An overheard cry had little 'probative value', the lawyer would say. That was for the crowd. And the men who were to judge this affair were military personnel, tough servicemen inured to plea and passion. What did they hear? What did they believe?

The examination by Julian Bevan stopped at 4.08 pm. Al-Matairi had to remain. It was time for his cross-examination.

3

TIM OWEN QC, CORPORAL DONALD Payne's barrister, was the first of the defence lawyers to address the witness. He began his cross-examination, gently lulling the witness into a placid state, outlining politely how his client had admitted abusing the detainees and now regretted his actions. Owen agreed with al-Matairi that he had suffered pain, he accepted that he was shouted at and had to adopt 'painful positions'. But the sense of sympathy that Owen portrayed wasn't as benign as it first sounded.

'I want to make clear,' he said, 'so that you're under no doubts about it, I will be suggesting that you have greatly exaggerated the amount of force used against you.'

The metaphor of drowning suddenly seemed apposite as the witness tried to draw breath, tensed his body to withstand the danger, and panicked. How could he respond to the provocation? If he had been coached he might have kept calm, he might have said, quite simply, that Owen was mistaken. But lawyers are not allowed to coach their witnesses in the British courts. It is against their code of ethics. Al-Matairi could only retort in a way that was familiar to him. And it instantly felt alien in this court in the middle of a British Army camp in the heart of Wiltshire. It had the scent of the bazaar, the coffee shop, where the theatrical, affronted reply to an unfriendly insult might be expected.

'I did *not* exaggerate. I told *less* than the fact that had happened. A *quarter* of what had happened. If you want . . . if you climb a mountain you would only . . . you are only the one to know what you feel and suffer. Nobody else can tell you.'

Owen ignored the grandiloquence.

'I want to make it clear that you have invented some incidents, said them for the first time today, things which never happened.'

'That is not possible,' said al-Matairi, and he embarked on a different allegorical foray. 'If you enter a restaurant and you see two of your friends eating together then maybe a year after that you would forget. But—'

'Yes, all right,' Owen interrupted. He asked al-Matairi to admit that he may have been confused, not least because he was hooded for most of his detention. The answer was chaotic.

'It was an insult both morally and psychologically . . . because . . . when I came here I just wanted to see your civilisation. I was taken to churches and to the British Museum . . .'

Before he could wander into further circumlocutions, Mr Justice McKinnon intervened. It was 4.30 pm and time to adjourn for the day. The judge had made it quite clear at the beginning of proceedings that he would stick to his timetable. They would resume the following morning.

Despite the brevity of questioning, though, Owen had managed to spread before the panel the entrails of his defence of Corporal Payne; yes, bad things were done, but not as evil as the prosecution or its witnesses claimed. Al-Matairi's fondness for nervous digression was unlikely to convince the army panel otherwise.

THE NEXT DAY OWEN RETURNED TO HIS sharp and direct style. He began by confirming that al-Matairi had made four statements to the RMP since the events of 14 and 15 September 2003. They were wonderful material for the pedant. In Owen's case, it was pedantry with a purpose: to discredit and unnerve. He pointed to a confusion in the witness' description of the soldier who had kicked him in the hotel and an officer he saw there at the same time. Who did he mean?

'I told them there were mistakes,' al-Matairi said. 'I told the investigator there were mistakes. If he was here I would know his face.'

'But Mr al-Matairi, you signed this statement. You *signed* it.'

'I didn't . . . I didn't read it all. I just looked at the last page and signed it.'

Owen asked about the witness' ability to identify any attacker in the detention block at Battle Group Main. Could he see with the sandbags over his head?

'With one bag I could see, but not in clear way. I would not know the face. I was shown a video, but I couldn't recognise any face.'

'Exactly.'

Mr Owen picked at the admission, the inability to identify any assailant. Despite seeing videos of Corporal Payne and privates Bentham and Slicker prepared by WO Spence in January 2004, al-Matairi had never been able to positively identify any of them. Owen was careful not to suggest the witness hadn't been hit at all. That would contradict the medical evidence. His purpose was to emphasise the lack of proven connection between the kicks and punches and Donald Payne. With that achieved, he advanced a new sequence of questions. They were close to ridicule.

'Your evidence yesterday was that you were beaten continuously for two days and one night?' Owen said.

'Yes.'

'Every minute of every hour, thirty-six hours, *continuously* beaten by many different people?' Owen couldn't keep the incredulity out of his voice.

'No, not when I was taken to the doctor. Not when I was taken to see the leader.'

'Quite correct. I accept that. Apart from that, *every* minute of *every* hour continuously beaten by different people?'

'All of us, not only me, apart from the last hours after Baha was taken.'

'You described yesterday being kicked with karate kicks. There's no description of this in any of your witness statements, is there?'

'That's no problem, I was tired.'

'Right. You said yesterday you think you were hit with a wooden stick.'

'Just once, a thin wooden stick.'

'There's no mention in any statement about being hit with a wooden stick.'

'No problem, I said that later.'

'You said yesterday that after you were kicked in the groin, you fell over, and you were lifted up and then punched again.'

'Yes.'

'No mention of that at all in any of the statements.'

'As you like.'

Owen had made his point and sat down. The questioning of the witness' honesty continued. Another barrister stood up to face him, Richard Ferguson QC, he of the 'hypnotic charm', LCpl Crowcroft's 'brief'.

Ferguson's initial tactic wasn't entirely obvious. He wanted to know whether al-Matairi had hired a firm of English solicitors to seek compensation for his treatment. Ferguson mentioned the name 'Mr Shiner'. Wasn't he acting for the witness?

'I do not know. I swear to God, this is the first time I hear this.'

The oath had the imprint of sensitivity, as though he should be ashamed of seeking redress and hiring a lawyer to represent him.

Ferguson asked if he had discussed claiming compensation with the other detainees.

'It is not time to talk about this,' al-Matairi said.

But it was, and Ferguson said Mr Shiner had written a letter to the Secretary of State for Defence claiming to be instructed on al-Matairi's behalf. How could that be?

Al-Matairi replied that he remembered signing something, someone had come to his house, although he had no idea what the document was about.

Ferguson then asked about al-Matairi's business. Al-Matairi *was* a businessman, wasn't he?

'Yes.'

'Interested in making money?'

Al-Matairi, an intelligent man, grasped the implication immediately.

'So that's it. The seriousness of this case is how we were tortured . . . how Baha died. That is the whole case,' he said.

'The seriousness of the case, Mr al-Matairi, is whether or not you are telling the truth . . . whether or not you were exaggerating the injuries which you received to make money . . . and one way of making money is to claim compensation.'

Ferguson's point was a tawdry one, based on a contradictory logic, but a powerful one nonetheless: if you are injured you have the right to sue for damages. If you do sue for damages then you clearly have an interest in exaggerating those injuries. Thus, the claim is tainted. Unfair as it might be, the logic can disrupt the way in which allegations are perceived. Mud sticks even if there is no mud.

Ferguson didn't dwell on the matter, though. Good cross-examination, it is taught, allows for packages of questions to be constructed. There is no need for a coherent narrative binding all the packages together. Following this rubric, Ferguson turned to another sensitive spot: the weapons found at the hotel. The pistols, the grenades, the bayonets, uniforms, a sniper scope. He listed them one by one and asked al-Matairi to explain why they were in his hotel. The answers weren't convincing. If he knew why the weapons were there then his explanations felt limp. If he didn't know, then his replies must have been fabrications, guesses at best, excuses. He couldn't win and he threw his hands up in despair.

'Please do not ask me about these things, I do not know about these things.'

It was enough for Ferguson. 'I suggest to you, Mr al-Matairi, that you are telling lies . . . that you are deliberately exaggerating whatever slight injuries you received.'

So it went on. Geoffrey Cox QC was next and like Owen and Ferguson before him, sensed a witness struggling on the stand. At 3.15 pm, an hour into al-Matairi's obvious torment, the judge declared a break. Mr Cox, with malice-spiked levity, protested.

'I was enjoying myself, sir,' he said.

The judge ignored the facetious remark and the witness was given ten minutes for tea.

WHEN THEY RETURNED, THE BAITING CONTINUED. EACH defence barrister was eager to repeat the accusations of exaggeration motivated by al-Matairi's supposed pursuit of money. Question after question after question, all with the same end in mind. There was another adjournment at 4.30 pm and they returned the next day to continue. It was relentless. After Mr Cox there was Jeremy Baker QC, and after him Jeremy Carter-Manning QC. The accused were entitled to the best defence they could muster and with a considerable slice of senior criminal advocates in the country representing them, that was exactly what they were receiving. Al-Matairi endured and finally was released from the court.

It had been a damaging experience; for the witness, for the victims who were to follow into the stand, and for the whole prosecution case.

4

THE NIGHT GUARD AT THE Hotel al-Haitham was the next witness to be called. It was now 4 October 2006 and the trial had already been going on for a couple of weeks. The witness was taken through the same events as al-Matairi: his humiliation at the hotel, in the lavatories, water flushed over him, his tears and praying. The weapons were mentioned. He replied they were for the defence of the hotel, protection from 'bad situations'. Then he was asked about his time at the detention centre, the punches, kicks, the sandbags. The memory became overwhelming and he began to weep.

'How can I calm down? I cannot calm down,' he said.

When he recovered, little things, ordinary things, things that reminded him of his own humanity, intervened. He was asked whether he ever knew what time of day it was in the detention block and he said he knew day had dawned when he'd heard the sparrows singing outside. Then he told the court about Baha Mousa. He'd heard him cry out several times that he was going to die; he'd heard cries of pain. And then there'd been a hiatus, the night guard said. The sandbags were removed as they entered the third day of their captivity. But the humiliation continued, even if the beatings stopped.

The night guard could only make one positive identification of his assailants. In January 2004 he'd been shown a video of an ID parade. He was able then to pick out one of the soldiers. That was the man whom he believed had been hitting and kicking him the most. It was Corporal Donald Payne.

Then cross-examination began.

Tim Owen was fairly reserved this time. His client had been positively identified and there was little mileage in trying to deny otherwise. After all, Payne had pleaded guilty to inhuman treatment. Owen asked a few questions intimating that others must have been at the detention centre causing harm to the detainees as well and this was not denied. There was little else for Owen to do.

Ferguson's approach was more invasive. He referred to a passage from one of the witness' later statements to the SIB. In that, the night guard had said that because he didn't know English he hadn't understood what had been shouted at him during his detention. He also said that he learned later the gist of what had been shouted because other detainees had told him. Ferguson asked whether the guard had ever discussed his evidence with anyone.

'Never,' the witness said.

Ferguson repeated the guard's statement. Surely he must have spoken about the case in order to learn what the guards had been saying? It was a small contradiction, but in the tradition of adversarial advocacy before the Bar in England, it was one to exploit.

'Who are the friends who told you and when did they tell you?' Ferguson asked.

'Where friends? When friends? What friends?'

'That's the point. What friends?'

'I didn't have friends when I was beaten.'

'But your statement mentions them.'

'I was . . . I was scared . . . I was so scared in there, even when I was giving evidence I was scared.'

'Yes, leave that to one side for the moment. Whatever your frame of mind you said "my friends have since told me". Did you say that or not?'

'Where did I say that?'

'In your first witness statement.'

'Yes, I told you, it is correct, yes. I said that, but I was scared.'

'Was this true what you said?'

'Yes, I told you. I said that because I was scared and confused.'

'But did you make this up?'

'Yes, correct. Being scared and . . . the humiliation I experienced.'

Ferguson knew better than to play on the inconsistency. It was obvious for all to hear.

The other defence advocates offered a light touch as well. Perhaps they recognised the night guard's emotional sincerity. Perhaps they were simply waiting for the more vulnerable witnesses to appear.

The hotel cleaner and part-time guard was next on to the stand. After giving his evidence in a composed manner, the defence attorneys circled. Julian Knowles, Owen's junior, followed the defence line that the witness was exaggerating. This served two purposes for Corporal Payne's defence.

It reinforced the argument that Payne hadn't killed Baha Mousa deliberately or negligently *and* it attempted to put the ill-treatment into the category of 'mild' abuse and only occasionally 'severe'. Payne had already pleaded guilty to inhuman treatment, but he hadn't been sentenced yet and wouldn't be until the end of the trial. If a sense of scepticism could be introduced about the extent of violence actually suffered by the detainees, then this might lessen the penalty imposed.

Knowles ran through the injuries that the cleaner had said he'd received. One particular allegation was questionable. The cleaner had claimed in a statement to his UK lawyer's representative in Iraq that he had seen a doctor after he had been eventually released from Camp Bucca. And he had said that although his kidneys were fine he'd suffered broken ribs. How did he know this?

'I was having pain. I was having pain for a long period of time. I couldn't touch my ribs.'

'Did a doctor ever diagnose you had broken ribs?'

'No. We didn't have X-rays.'

With remarkable swiftness, Knowles finished his questioning. He had secured the information he needed. It was a very small piece in the tapestry the defence was weaving and it was stored away for closing submissions.

Richard Ferguson QC, representing LCpl Crowcroft, the soldier who along with Kingsman Fallon guarded the detainees for most of the first day after their arrest, pursued a different approach. He wanted it to look as though the violence directed against the detainees was limited whilst his client had been watching over them. He wanted the panel to think the real abuse began *after* Crowcroft and Fallon handed over guard duty to A Company at about sundown on the first day. He even wanted to suggest that there was compassion amidst the brutality. He didn't succeed.

Ferguson said, 'On occasions, when you were provided with water it would be with kindness, wouldn't it?'

'But he would give me *hot* water. He would put the bottle to my mouth, but the water was hot . . . unbearable.'

Geoffrey Cox also wanted to suggest that most of the violence took place after his client, Kingsman Fallon, had left the detention block.

'There's always been a clear distinction in your mind between the level of severity of the violence on the second night and the first day, hasn't there?' Cox asked.

'The second night it was really severe. I was beaten for long periods. On the first day I was hit when I couldn't maintain the stress position.'

Cox asked about the weapons. Being a part-time guard as well as cleaner, the witness might help understand why the guns were found at the hotel. He confirmed that he knew about the rifles, and would have used them for protection. In any other context this might have been suspicious. In a lawless city where violent theft, vindictive attacks and hijacks were commonplace, the possession of weapons was a sensible precaution. The cleaner even explained why magazine clips were taped together, as one might see in war movies. All Iraqis had to undergo military service and this was how they'd been shown to look after their ammunition. There was nothing sinister in it.

Like the night guard, the cleaner's ordeal ended abruptly. He was released from the witness box and sent home. Another sharp introduction to the process of British justice.

5

THREE DETAINEES FOLLOWED. THE MAN who serviced the generator at the hotel recalled how he saw a soldier stamp on Baha Mousa when they were arrested. It was in the hotel lobby, he said. But he didn't really know what happened. He had seen no injuries to Baha's face afterwards. Then the scene shifted and they were at the detention block. 'Torture started,' the generator man said.

'Torture?'

'They made us stand by the wall our hands stretched out, our knees bent; half sitting half standing . . . kicks on my thighs, punches on my face. He would strangle me to make me stand. A headlock.'

Other soldiers were there too. They were punching and hitting. At night-time the 'torture' got worse. First day became second day. The beatings continued, but were more intense. There were lots of soldiers. He saw them when his hood was lifted to take a drink of water. The beatings had no reason. Then a metal bar appeared. He said he was hit with it. Other blows, kicks, continued. So hard he collapsed. He said he screamed so much they brought a stretcher and he was taken to the clinic at the base. He saw a medic and told him he'd been hit. The medic just shook his head and gave him an injection.

The cross-examination of the generator man was light, almost polite. His evidence had been consistent, solid. Suggestions were made that he couldn't be sure about when particular attacks occurred or who might have inflicted particular injuries. This was generic questioning. It applied to all the detainees. The fact that, by their own evidence, they had sandbags encasing their heads, were exhausted and hungry, and the events they were describing happened over three years before, all emphasised an inherent uncertainty coursing through their testimony. An astute observer in the courtroom would know the purpose of these seemingly innocuous questions; where there were no obvious contradictions, the ground was being prepared for submissions that their evidence was inaccurate, or just mistaken. It didn't really matter how these witnesses appeared. They would be damned by an argument, not evidence.

The generator operator was dismissed and the man in charge of the restaurant at the Hotel al-Haitham appeared. He was the one who'd tried to help the soldiers in the hotel locate one of the owners, the one who had run away. The soldiers asked him to take them to the man's house, which he did. But he wasn't there and the soldiers weren't happy, he said. He was hit across the face when he asked for water. Then he was taken to the detention centre, the place where the beatings began: stress positions, kicks to the kidneys when he failed to stand/kneel/squat as instructed. It was a familiar refrain. He added the 'dance like Michael Jackson' account. Soldiers made him dance the morning after Baha died. From Camp Bucca he was taken to the hospital at Shaibah where he stayed for two weeks. Then he was back at Camp Bucca for another couple of months.

There was one more thing he wanted to say. Corporal Payne, whom he identified in a video, was the 'one who killed Baha . . . he was shouting all the time . . . he would beat us and then go away and then come back and beat us'. Even with a sack over his head, he was able to see Payne through the small holes.

Sensing a slight fissure in the man's testimony, Julian Knowles, acting for Payne, stepped in.

'You didn't see what happened to Baha Mousa, did you?'

'No.'

'So who told you it was this soldier who killed Baha Mousa? Who have you been speaking to?'

'That soldier came, Baha was taken to another room, the soldier was shouting, it gave me the impression that it was him who killed Baha.'

'So it's just a guess on your part . . . there were lots of soldiers around . . . you have no idea, have you, who was doing what to whom . . . you had sacks over your head.'

'Correct.'

'Have you spoken to other people about what happened?'

'Of course . . . my family.'

'Other detainees?'

'We weren't allowed to talk to each other.'

'What about over the last three years?'

'Yes.'

The suggestion of collusion was enough.

Then it was the turn of Ahmed Maitham, the car-crash detainee. He had nothing to say about events at the hotel. And, as it was common ground that he only arrived in the detention block at Battle Group Main late on the evening of the day the other detainees were arrested, his evidence was dealt with quickly. Two of the defendants couldn't have been involved in his treatment. Crowcroft and Fallon had left the facility by that time. Whoever had assaulted Maitham it wasn't them. Within an hour the witness was released. 3,000 miles. It was a long way to come for sixty minutes of painful recollection.

So this was the state of the detainees' testimony. Would it have been enough for the military panel of jurors to appreciate the feelings of these men subjected to such vindictive and casual violence? What could those colonels and majors have heard amidst the legal bluster? Had they sensed the pain and anguish? Or had they doubted it? Had the persistent accusations of exaggeration or lack of identification or uncertainty in timings and events or contradictory accounts introduced disbelief?

6

THIS WAS THE EXPERIENCE OF the detainees. One by one:
They were sworn in.
They were asked questions.
They were encouraged to recall the details of those events three years before.

They were given no assistance.

They were expected to remember.

They were questioned by the prosecution.

They were cross-examined by the rank of defence lawyers.

They were the players in a ritual developed over centuries, a ritual they had never experienced before.

They had already given statements to the Royal Military Police, several times over in some cases, telling and retelling what they had experienced.

They couldn't speak out of turn.

They couldn't relate their account by referring to any notes.

They couldn't make speeches.

They couldn't protest.

They couldn't lament.

They couldn't make claims.

They couldn't argue.

They couldn't digress.

They had to perform like actors on a stage, but with more conviction. One could say they were abused afresh.

7

AFTER THE DETAINEES HAD BEEN heard, the trial surrendered to a remarkable deceleration. There were interruptions, digressions. The *Daily Mirror* published a picture of Colonel Mendonça, his face partially blacked out, but not wholly obscured. It was in contravention of an order given right at the beginning of the court martial. Photographs of the accused soldiers weren't supposed to appear in the press. It was for their security and safety, a rare privilege for a defendant.

Colonel Mendonça's lawyers objected. The *Mirror*'s lawyer and editor were commanded to appear before the judge to apologise, to explain their error, and to guarantee that it wouldn't happen again. It was humiliating, but did they care?

Then there was another interjection. Julian Bevan QC, leading counsel for the prosecution, had something to confess. He was in the car park of the Court Martial Centre. A man spoke to him. They chatted for a few minutes about Afghanistan. It was trivial, idle, polite conversation. Two oldish men sharing a few comments. Except that the man whom

Bevan had spoken to was one of the panel of military jurors. And he wasn't supposed to have any communication with them. It was against the rules. Bevan had to explain that he hadn't recognised the panellist. He had to apologise for his unintentional mistake. No one seemed to mind, but the etiquette was observed, another unnecessary interlude in proceedings, and for those observing, something of a whimsical one at that.

And as if to irritate the timetablers and courtroom managers, one of the accused, Kingsman Fallon, was struck down by appendicitis and had to be taken to hospital.

Each time there was an interruption the trial was halted, the timetable revised and the progress of the case deferred. It was all part of the inexorable demands of a fair trial in the British legal system.

8

FEW DOUBTED THAT IT HAD been a poor beginning for the prosecution. With the fierce talents of the assembled defence QCs ranged against them this might have been predicted. But there were solid witnesses to come, or so it must have been hoped.

In the temporary absence of the remaining two detainee witnesses, the father and younger son (who had yet to be transported to Britain), Captain (now Major) Moutarde was called to the stand. It had been agreed to slip him in, ahead of his slot in the list of witnesses.

Major Moutarde was the adjutant for 1QLR and in charge of unit discipline. Those representing Major Peebles and the colonel, both accused of neglect of duty, were interested in his evidence. They might have been asking why wasn't Major Moutarde charged along with their clients? Didn't he share some of the responsibility for the system in operation at Battle Group Main?

It was a fair question. And if Moutarde, then why not Lieutenant Rodgers? He was the Company commander and the SIB knew he'd visited the detention block on a number of occasions. And if Rodgers, why not the second in command, Major Süss-Francksen? Surely he bore some responsibility too? If this reasoning continued, most of the officer corps of 1QLR should have been hauled before the court.

With Major Moutarde ready, Mr Bevan asked his questions. All he wished to establish was the layout of Battle Group Main. It was an oddly subdued role for an officer who held the vital administrative position of Battalion adjutant. Surely he knew more than the geography of the base?

Most of the defence advocates passed on the chance to cross-examine; there was nothing contentious here. Mr Langdale, acting for Colonel Mendonça, thought differently. He smelt an opportunity. Major Moutarde might prove useful for his client's defence. He could testify to the character of Colonel Mendonça.

'Would it be right to say', Langdale asked, 'that Colonel Mendonça made considerable efforts to bring 1QLR up to speed with regard to the deployment to Iraq?'

'Yes, it would,' said Major Moutarde.

Langdale lauded the achievements and high standards successfully imbued into the regiment because of his client's professionalism. Moutarde was happy to echo the praise. Langdale stressed the standing orders issued concerning the treatment of detainees and in particular their transfer to Camp Bucca within fourteen hours of arrest. Was this something the CO had been keen on respecting?

'Very much so and wherever possible,' said Moutarde.

And what about Major Royce? He had been Major Peebles' predecessor in the role of Battle Group internment review officer, the man responsible for dealing with detainees.

Langdale asked, 'Once Major Royce had been appointed as BGIRO, did it become clear that what Major Royce had passed on as being information that he had received from Brigade was that hoods and restraints were approved?'

'Yes.'

'It was a verbal instruction which in due course became a standard operating procedure?'

'Yes.'

This was a moment of great significance. Langdale asked him whether he was aware that stress positions were allowed to be imposed on detainees in 1QLR.

'Yes, sir.'

Was he aware that stress positions were used in the detention facility?

'Yes I was, sir.'

They were used to condition someone for tactical questioning, he said.

But it was for a limited period. Not prolonged. Besides that, he never saw any abuse of a detainee. If he had he would have reported it. He felt sure all the officers would have done the same.

'I think the officers were almost without exception high quality officers . . . not just in terms of performance but in terms of understanding the mission, their own personal example and moral courage,' Major Moutarde said.

The transformation from prosecution filler to defender of the battalion's honour must have sickened Mr Bevan. Major Moutarde had become a character witness for Colonel Mendonça, a character witness for the whole unit. His evidence supported the proposition that Mendonça had presided over a professional and ethical outfit which followed the rules laid down by Brigade with vigour. There may have been ill-treatment. Indeed, there was no denying it. If it had occurred it was out of character and the result of one, maybe a few, rogue soldiers, soldiers who had lost their bearings, who had deliberately ignored established procedure. And if that procedure was in itself against proper standards of humane treatment, then that wasn't the fault of the battalion. It was the army's responsibility. Brigade had set the conditions. The battalion just followed orders. If hooding and stress positions were illegal, well, that was precisely what Brigade had required through its 'conditioning', its 'shock of capture', its harsh interrogation regime. This was the tenor now of Major Moutarde's evidence.

'In general,' Moutarde said, 'the whole issue of moral courage was strong in the battalion . . . every officer from the youngest second lieutenant up to the commanding officer displayed moral courage every day.'

Langdale mentioned the matter of the stolen money at the Hotel al-Haitham. What was Colonel Mendonça's reaction to this?

'He was livid . . . he spoke to the Battle Group and told them they were there to help the people of Basra, not steal from them.' The CO had come down hard on the men. He wasn't going to stand for such behaviour.

So was Colonel Mendonça's response when faced by something unethical, something that would harm their mission in Basra, to react instantly? Absolutely, Major Moutarde said. So how likely, then, would it have been for Colonel Mendonça to ignore obvious cruel ill-treatment of Iraqi civilians? Not likely at all.

One final question.

'What would you say to the suggestion that Mendonça was somebody who was likely to be in neglect of his duty?'

'I would say it was rubbish and I wish I could use a word with two Ls in it, but I don't think it's appropriate for this courtroom.'

Major Moutarde's vivid support for Colonel Mendonça was extraordinary.

Bevan had to retrieve the prosecutorial initiative. He asked to re-examine the witness and began by enquiring whether Moutarde and Mendonça were friends. Yes, was the answer. But this was too feeble an attack; the men had been officers together for some time. It would be surprising if they weren't friends.

Bevan changed tack, groping for a weakness to undermine Moutarde's testimony. He asked about Moutarde's direct knowledge of the treatment of the detainees. It was known that he had visited the detention block. What had he seen?

'Were the detainees hooded when you visited the facility?'

'Some were, some weren't.'

'Were stress positions still being used on the Monday?'

'I can't remember.'

'Were they cuffed?'

'Again, I can't remember, but I presume so . . . for security.'

What about 'conditioning', about training? Moutarde couldn't help. It wasn't his responsibility. He didn't know any details.

What training did the private soldiers have in maintaining stress positions, in hooding, in conditioning techniques?

'I don't know,' Moutarde said.

Bevan became more aggressive, his questions increasingly poisonous.

'Help me on a training point – how do you force a person to maintain a stress position and at the same time act humanely? . . . You were the adjutant, you help me, you knew this was going on – what's the limited time you can properly put a person in the stress position for, in your view as a professional soldier?'

'Um, I don't know.'

'Five minutes?'

'Possibly.'

'Half an hour?'

'Possibly.'

'An hour?'

'That might be slightly too long . . . without a break.'

'Two hours? What do you say?'

'I say I don't know.'

'Do you recall ever seeing a document that sets out the procedure for the treatment of detainees?'

'I can't recall seeing one. I don't know.'

'Any document for monitoring or checking the detainees weren't ill-treated?'

'Not that I can remember.'

'There was no procedure, was there? No written procedure.'

'Not that I recall.'

'You visited the detention facility occasionally?'

'Twice a day.'

'Were you under orders to do that, or did you do it of your own volition?'

'My own volition.'

'So there was no procedure for you to do this?'

'No.'

'Was there any order for any officer to visit the detainees to check their welfare?'

'Aside from being a function of command and individual moral responsibility, no.'

Bevan decided to show the video of the detention facility and Corporal Payne berating the detainees. It was a shock tactic, to draw some flicker of moral doubt into Major Moutarde's insouciance.

Bevan asked, 'Do you find Corporal Payne's conduct untoward or not on that video?'

'No. It was in line with the directions on conditioning for limited periods of time to prepare them for questioning.'

'Are you telling this court that you regard Corporal Payne's treatment humane?'

'If done for a limited period of time as part of the preparation for tactical questioning, then yes.'

'You heard the language used . . . Corporal Payne's words . . . "Shut up you fucking animal." Was that untoward?'

'If I'd heard that word I may have investigated, but not necessarily intervened.'

But he'd never heard that language used during his visits or in passing, he said.

Bevan kept challenging the moral assurance of Moutarde and his evaluation of 1QLR and its officers. He returned to the stress position: was one of its objectives to cause pain or suffering? Was that the purpose?

'As I understand it the purpose was to induce some discomfort,' said Major Moutarde.

'Discomfort would embrace pain?' And Bevan spelt it out so that there was no confusion: 'P-A-I-N.'

'Of course . . . some semblance of pain.'

'Is that in accordance with the Geneva Convention? In your view as adjutant?'

'In my view, yes.'

'These people were prisoners.'

'They were detainees, yes.'

'They weren't there voluntarily.'

'No.'

'And you have an obligation to treat a prisoner . . .'

It had now become an aggressive interrogation. Bevan was impassioned to such an extent that Mr Langdale intervened.

'My Lord,' he said wearily, 'I've held back for quite a long time . . . Prosecution counsel are not allowed to cross-examine their own witness either directly or by the tone of voice. This is all basic stuff. We're reaching the stage where Major Moutarde is being cross-examined.'

Bevan was rebuked but continued nonetheless. He asked Moutarde about his visit to the detention facility on the Monday. Was he satisfied with what he saw?

'Yes . . . it wasn't tiddlywinks . . . people were being conditioned. So it didn't look like a scene from Butlins.'

Langdale stood up to object again. He protested about the grossly leading questions. It was all against the rules.

The judge admonished Bevan once more.

'I'll go and have lunch alone,' Bevan muttered.

Little should be made of a joke. In the context of the trial, however, it signified much. From whom was he exiling himself? The club of QCs and junior barristers and solicitors? It certainly wasn't the defendants, nor was it the witnesses. That would be against the rules. But lawyers lunching together? That would be acceptable, perhaps even encouraged. It displayed the professional ability to separate the courtroom battle from the humanity of normal social interaction. They were to be trusted not to divulge secrets or concoct deals which might undermine the whole process of law. That was the ethical exoskeleton keeping them secure in their privilege. It didn't mean to be exclusionary, but it was exclusive. For those looking from different cultures and worlds and professions, it must have been disconcerting.

Lunch was taken.

9

A SENSE OF BETRAYAL PERMEATED the atmosphere in Bulford Court Martial Centre. There was a feeling of the case slipping away. Witnesses were either being undermined under cross-examination or they were proving valuable assets for the defence. Bevan had tried to recapture some of the moral ground, challenging the easy way in which his witness, Major Moutarde, could applaud the command of Colonel Mendonça. It meant overstepping the professional mark, necessary if a little desperate. He couldn't let the adjutant walk away from the witness box without being made to feel shame at what had happened under 1QLR's watch. The video of Payne in the detention facility had been intended to show the falsity of the claim that all the officers of the battalion were fine specimens of moral courage. Even that tactic had backfired. The video hadn't elicited an admission of horror from Major Moutarde. The exact opposite had occurred. The conditioning that they all witnessed was entirely acceptable to him; it did nothing more than follow established interrogation protocol as far as he was concerned. The attempt to highlight the obscene language was ill-fated too. For who in the armed forces would really be shocked by the swearing screamed by Corporal Payne caught on camera? Most likely, they would have been on the receiving end of such language every day. That was the nature of things in the army: tough, uncouth, coarse. To think otherwise was naïve. In any case, did it really invoke a sense of a war crime? Or inhuman treatment? Or even serious neglect of duty?

The need to restore the moral purpose of the prosecution was acute.

SAC Scott Hughes was called on 26 October 2006. He was the young man who had galvanised the investigation back in September 2003 by providing to the RMP a clear, seemingly unprejudiced account of the casual, open and systematic violence used against the detainees.

Now, as Hughes marched into the witness box he appeared subdued. He looked *very* young, younger than his twenty-odd years. As soon as he began to answer questions it became clear that his recollection of events three years previously was seriously impaired. He couldn't remember much of what he had put in his previous statements to the RMP. When he was asked about wandering into the detention facility whilst accompanying the *GMTV* crew on the second day of the detainees' incarceration, Mr Bevan had to help him, he had to be referred constantly to the statements he gave to the SIB at the time. Hughes' memory failed time and again. The only episode he could recall with any clarity was the 'choir', when Corporal Payne administered his macabre game of singalong by hitting the detainees in the gut or the kidneys in order to extract a grunt, a noise.

Tim Owen, acting for Corporal Payne, wished to challenge the accuracy of Hughes' evidence, but he didn't want to suggest he was lying. He wanted to plant the idea that maybe Hughes had a reason to exaggerate what he had seen.

'Did you get on with Major Greenwood?' Mr Owen asked, referring to Hughes' superior officer in media operations.

'Yes, sir.'

'Nice man?'

'Very nice man.'

'Easy to talk to?'

'Yes, sir.'

'Approachable?'

'Yes, sir.'

'Not formal, distant, cold, aloof?'

'No, he was a spot-on boss, sir.'

'So why didn't you raise what you'd seen in the detention centre with him?'

'Um, I . . . I'm not totally sure, sir.'

'Not totally sure. Think about it,' and Owen began to list the opportunities Hughes had had to speak to Major Greenwood *after* the journalists had been dropped off.

'Can you think of *any* difficulty, *any* logistical difficulty, *any* impediment, *any* reason why you couldn't have raised with Major Greenwood on the 15th what you had seen?'

'No.'

'There is no good reason, is there?'

'I don't know, sir.'

'If everything that you've told us is true, then this was a serious incident you witnessed?'

'Yes, sir.'

'Vivid?'

'Yes, sir.'

'Horrible?'

'Yes, sir.'

'Shocking?'

'Yes, sir.'

'Why did you not tell or raise it with Major Greenwood as soon as the journalists had disappeared on the 15th?'

'I don't know, sir.'

It sounded implausible. Owen could understand why Hughes might not have wanted to mention anything when the *GMTV* people were there. That would have been difficult, perhaps incendiary. But once they had gone? The lack of an explanation left a void which Owen was quite capable of filling with a damning conjecture: SAC Hughes wanted to get out of Iraq. He was afraid. He didn't like it there. When he heard of the death at Battle Group Main, he saw an opportunity to have himself airlifted away, back home.

'You were desperate to get out of Iraq, weren't you?' Owen said.

'I wouldn't say I was desperate. I wanted to go on my rest and recuperation.'

'Think again, Mr Hughes. You were desperate to get out of Iraq, were you not?'

'I wouldn't say I was desperate.'

'You made it clear to those in charge of the media ops unit that you felt you couldn't cope with life in Iraq after this incident.'

'I wanted a rest. I can agree that. But I wouldn't say I was desperate to get out.'

'But you'd only been there six weeks. Let me put it another way: did this incident in fact enable you to get out of Iraq?'

'Yes, sir.'

'Apart from the identification from the back of WO Spence's blacked-out Land Rover you didn't come back to Iraq, did you?'

'No, sir.'

'The reason you didn't come back to Iraq was, you said, because you were stressed and distressed by what you'd seen at QLR Main and you said you were afraid?'

'No. I don't think I said I was afraid, sir, no.'

Hughes had only come forward *after* he had heard one of the detainees had died. The implication was he'd seen an opportunity, a chance to get out of that hellhole. But this defence theory wasn't entirely self-serving. In order to take advantage of events Hughes still had to have seen something, some violence, some ill-treatment. And he had been one of the first to talk about the 'choir', that by now emblematic malignity of the sickening trough of corruption pervading Battle Group Main. Was it enough to claim Hughes was exaggerating?

Owen's approach was subtly generous. He accepted that the 'choir' had occurred and Hughes must have seen it. Corporal Payne had admitted the 'choir thing' as Owen now called it, as though it was somehow disconnected from any human construction. But Owen suggested that this was the *only* abuse Hughes had seen that day. Everything else was an exaggeration. The choir was just one of those isolated, ill-conceived incidents which Hughes had happened to see. It represented nothing more than that. And to back up this theory Owen pointed to Hughes' inability to remember any of the details of what he had seen *other than* the 'choir thing'. He took Hughes through a line of reasoning that a child could appreciate.

'I suggest that the reason that is the one thing you can remember is that it's practically the *only* thing that actually happened that you saw that day. That's why you can remember that one, because you actually *did* see it, do you follow?'

'I see what you're saying, but I think you're wrong.'

'You see, if somebody, any of us, has witnessed an incident that has really happened, that we have *really* seen, then, of course, we may forget details, we may forget the precise sequence of events. But if we've in fact witnessed something, it all comes back once you've started to put it together or you've read one section of the statement to really refresh your memory. Do you understand what I mean?'

'I can see what you're saying, sir, yes.'

'But if in fact you have recorded in this statement a whole load of exaggerated untrue allegations to spice up this account, then, of course, when you read the statement your memory isn't going to be refreshed. Because you never actually saw it. Do you follow?'

'I see what you're saying, sir.'

'I suggest, Aircraftman Hughes, that is what you've done in giving evidence today. Your memory wasn't really refreshed by looking at that statement, was it?'

'Bits were, yes.'

Owen's was a persuasive argument. Would memory of such an ugly incident have been erased by the intervening three years? Would one's grasp of the sequence and nature of violence have been so tarnished, even lost? Perhaps the matter isn't as simple as it might appear, though. The psychology of memory remains a science in flux. Some accounts hold that the ability to recall events is always subject to interference. The banal incidents of life can disappear within seconds. Their very ordinariness pushes them away almost as soon as experienced. Less trite events are unlikely to be erased so easily and remarkable ones have a good chance of being stored in one's long-term memory.

But that's only a theory. The whole process of recall can be affected by multiple factors. Original perception is subject to a host of distorting or enlightening influences: the length of time a witness saw the events unfold, the level of stress experienced by the watcher, the impact of psychological arousal (up to a point of trauma), heightened perhaps by violence. There is even a theory of 'flashbulb' memory, where a highly unusual happening prompts a camera-like fixing of detail. Then there are those factors which can interfere with later recollection: the time elapsed since the original event, the frequency of remembering, details added by others. And the imposition of stress when attempting recall, such as cross-examination by an aggressive barrister in a courtroom full of scrutinising press and soldiers and lawyers, can trim memory, even prevent its retrieval altogether.

Did this make Owen's argument flawed? No. Not because it took little account of psychological research or theory. Who was there to explain all that anyway? It was because Owen projected a common-sense notion that had nothing to do with science. It appealed to the jury members at a raw emotional level. Each one would ask himself 'Would I remember the detail?' And the likely reply of those major generals and colonels would be 'Yes.'

The practice of law may seem scientific, but it relies heavily upon primitive perceptions.

10

So much for memory, but what about wilful amnesia?

October 2006. It was time for members of the Rodgers multiple to be called. Private (now Kingsman) Christopher Allibone entered the witness box first. And instantly complained, sounding like a disgruntled schoolboy reluctant to take an exam. He hadn't been given the chance to read his previous statements, he said. He couldn't remember much without them. He was nervous and had difficulty reading. And he suffered from a stutter.

The judge adjourned the hearing so that the soldier could have his statements read to him. A witness support officer stood by Allibone's side when he returned to the box to help him read the statement when called upon to do so by the advocates. But the lines of communication were strained. The judge often couldn't hear what Allibone was saying. There was a whiff of the music hall about the exchanges.

Prosecutor: 'Can you help us with who was there when you were deciding who should do the stags?'

Allibone: 'Everyone in the multiple.'

Prosecutor: 'All the multiple? Including Lieutenant Rodgers?'

Allibone: 'I'm not sure. I don't think he was there.'

Prosecutor: 'Did the stags include all the multiple?'

Allibone: 'I think it was just the privates.'

Judge: 'Just the what?'

Allibone: 'The privates.'

Judge: 'Just the four of you?'

Prosecutor: 'The privates.'

Judge: 'Just the privates, thank you.'

Prosecutor: 'Did you see Corporal Payne at that stage?'

Allibone: 'Yes.'

Prosecutor: 'What was he doing?'

Allibone: 'I'm not sure.'

Prosecutor: 'Did you speak to him?'

Allibone: 'He told us where the water was if anyone needed it, he told us where certain important locations were. Whoever needed the med centre, where that was . . .'

Prosecutor: 'The medical centre?'

Mr Ferguson: (*interjecting with irritation on behalf of LCpl Crowcroft*): 'I'm sorry, I can't hear, My Lord.'

Judge: 'You'll have to say that again. You said "Where the water was if anyone needed water. The toilets were . . ."'

Allibone: 'I didn't mention anything about toilets.'

Judge (*quizzically looking at the witness*): 'You didn't mention anything about toilets?'

Prosecutor: 'We'll come on to the toilets later.'

Ferguson complained again that it was impossible to follow what the witness was saying. Allibone was mumbling. He was incoherent. His answers were confusing. He was running his words into one, long, incomprehensible slur. Then as though with a momentary clearing in the mist Allibone remembered Corporal Payne telling him to punch the detainees in the stomach if they didn't maintain the stress positions. That would avoid bruising, he had been told. And he said Payne kicked the detainees. It was a side kick, Allibone said. A 'side kick?' the judge queried. Allibone was requested to demonstrate what he meant.

'This is embarrassing,' said Allibone, but he came out of the witness box, taking centre stage in the courtroom and executed the karate-style kick anyway. Prosecuting counsel interpreted his actions, translating the movements for the benefit of the stenographer: 'Right leg side kick, bending the knee and kicking out at hip height.'

'Before you sit down,' the prosecutor said, 'were there any other types of kicks?'

'Do you want me to . . . ?'

'Yes please.'

And in a bizarre exhibition, Allibone showed how to perform the roundhouse kick, spinning round and kicking into an imaginary midriff. He was an expert in karate, he told the court. And kung fu. And Thai boxing. And a number of other martial arts.

With that over, he was asked about his conduct that Sunday night of the detainees' arrest and the following Monday. Allibone said he pushed some of the detainees, maintained the stress positions as ordered by Payne and kept the prisoners awake with the iron bar banged against the floor.

This admission caused Geoffrey Cox, who represented Kingsman Fallon, to erupt into action. He stood and poured out his disdain for the prosecution's case. His fellow defence attorneys joined in. They were incensed, very theatrically.

Cox called the proceedings 'surreal . . . Lewis Carroll territory'. He said the witness should be warned that his evidence might incriminate him. And then he launched into a tirade: his client, Fallon, had been charged with inhuman treatment for the very same actions which Kingsman Allibone was now saying he undertook as well. Cox pointed out that the detainees who had given evidence told how they were abused continuously from the Sunday until the Tuesday morning, that Fallon and Crowcroft were only in the detention facility on the Sunday afternoon, that the ill-treatment continued *after* they left their shift, that Corporal Payne wasn't the only perpetrator of violence (as Hughes testified), and that therefore others in the Rodgers multiple must have been involved in inhuman treatment. And yet it was only Fallon and Crowcroft who were being hauled up before a court martial. This was 'double standards', Cox said. He had a point.

Tim Owen was even more vociferous in condemning the 'arbitrary, inconsistent, irrational, frankly absurd approach the Crown have taken to picking and choosing a few people to take the rap'. Had the witnesses been granted some kind of secret immunity? he wondered out loud.

Absolutely not, said Mr Bevan. He explained why the prosecution had targeted Fallon and Crowcroft: they were in the detention facility for a considerable period of time; they were seen by Private Lee assaulting the prisoners and, Bevan said, they bragged about their exploits after the event. That separated them from the rest of the guard. In Payne's case, there was evidence of gratuitous violence that was of a different order from the force used to maintain conditioning. As for Allibone, Bevan said he had considered the admitted violence inflicted on the detainees, but on balance had concluded that it didn't amount to inhuman treatment.

It was a fine line to draw. Why, indeed, had so few been prosecuted? Why Fallon and Crowcroft and not all those others involved in subjecting the detainees to a long period of abuse? Why had no one from the Rodgers multiple been charged? What was the sense in only selecting a few when so many by their own admission had been involved? All that investigation for so little return. What had happened? Of course, if everyone involved in guarding the detainees, keeping them awake and maintaining stress positions, had been prosecuted, who would have been presented as witnesses for the prosecution? There would have been the detainees, but they had been hooded. And there would have been Scott Hughes and the other media drivers, but they had only been in the centre for a maximum of an hour or so. If the Rodgers multiple were left out, the case would have seemed very thin.

The spat between lawyers, for so it appeared, continued for only a short time and then subsided, as though the fighters were exercising their sparring talents, but holding themselves back for the main bout to be staged later. All these arguments would have to be repeated when the prosecution finished presenting their evidence.

In the meantime, it was back to Kingsman Allibone and his cross-examination. The lawyers had fecund territory to inhabit with all those bizarre martial-arts displays and incomprehensible mutterings.

Corporal Payne's barrister accused Allibone of being the one who had used karate-type kicks. He accused him of trying to pin all the blame on Corporal Payne. And finally Allibone was asked 'Do you know what perjury is?'

'No,' Allibone said.

'It's probably just as well,' said Tim Owen in well-rehearsed pantomime style.

11

THE INTERPLAY BETWEEN BARRISTERS, BETWEEN barristers and witnesses, between witnesses and judge was becoming increasingly tetchy, strained, as though the protracted proceedings had induced a sense of familiarity breeding collective and mutual disrespect. Perhaps they all felt oppressed by a burgeoning miasma of futility. But amnesia of events in September 2003 was threatening to become endemic. Private Appleby, the next soldier witness, couldn't remember. Private MacKenzie couldn't remember. Their testimony repeated the same admission time and again: 'I don't remember', 'I can't remember', phrases like a chorus in a perpetual song. The only thing they could recall was the pre-eminent role of Donald Payne. In his case, the fog was lifted again and his brutality highlighted in their memories.

Both were vulnerable. Private MacKenzie had kept a diary during his tour of duty in Iraq, a book which had been confiscated by the prosecution and revealed as part of the evidential package. The junior barrister acting for Corporal Payne, Julian Knowles, seized upon its contents to discredit MacKenzie's character and therefore his already limited credibility. And in truth it was a treasure chest for the cross-examiner.

Knowles read out some of the extracts. It wasn't very pleasant.

"'Public order training for police,'" Knowles read, "'Women with leather faces pushing to claim pensions . . . punched a policeman for not doing as told . . . sore feet and very tired . . . kids throwing stones, hit on calf and shoulder, no damage . . .'"

MacKenzie couldn't remember hitting anyone, let alone a policeman.

Knowles interrupted his own recitation of the diary and asked MacKenzie about A Company's nickname. 'The multiple was known as the Grim Reapers, is that right?'

'Yes,' MacKenzie said.

'Why was it called that?'

'I believe it was because it was our multiple that got the first kill in Iraq.'

'Sort of badge of honour for killing people?'

'I don't know.'

Returning to the diary, Knowles read out an entry that said Lieutenant Rodgers had been punched by an Iraqi civilian and the multiple 'filled in' the assailant: "'he was battered from head to toe so we let him go . . .'" But MacKenzie could remember nothing about such an incident. The fog had descended once more.

Knowles read on. "'Caught three Ali Babas. Beat the fuck out of them on the back of the Saxon . . .'"

Again, no recollection for MacKenzie. Then the diary entries referred to the detention of the men from the Hotel al-Haitham. The word 'conditioning' was mentioned and the 'fat bastard' who kept removing his hood and that it was this man who 'stopped breathing . . . we could not revive him. What a shame.'

Knowles wanted to know whether this was a sarcastic comment.

'Yes,' said MacKenzie.

It was a moment of candid callousness; in the recorded entry and the admission now under oath.

After pausing to allow the import of this attitude to seep into the whole atmosphere of the trial, Knowles pressed on and discovered that MacKenzie couldn't recall being questioned on his diary entries by the SIB. Should he have been prosecuted for the diary's contents? Knowles asked. MacKenzie didn't know.

And then he was asked about the *Daily Mirror* photographs, those sold to Piers Morgan and his journalists in 2004 which were proven to be false. Was he the person in the pictures, shown urinating on the head of a prisoner?

MacKenzie was advised that he had the right not to answer such questions. Any admissions could be used against him. He took the advice and refused to answer on the grounds that it could incriminate him. And he refused to answer the next ten questions all related to the *Daily Mirror* story on the same grounds. Knowles accused him of being violent and dishonest. Unsurprisingly, MacKenzie denied it. It was as though the defence had transmuted into the prosecution, eviscerating another witness.

There was worse to come.

12

FOR A MOMENT, THOUGH, THE news was all centred on another trial, far away, but intimately connected to the court martial. Saddam Hussein had been tried in the Special Iraqi Tribunal and confronted with various crimes committed within Iraq during his reign. The evidence of genocide and crimes against humanity had been presented and yet somehow it had been the theatrical elements of Saddam's trial that had captured the media's attention: the histrionic outbursts, the shouting, the sheer black comedy of such performances. Those horrific facts revealed by the American-backed prosecution failed to attract as much interest as the erratic behaviour of its participants. It was the trial that was the 'show'. And on 5 November 2006, just as Private Aspinall was called into the witness box in Bulford Court Martial Centre, the tribunal in Iraq announced its sentence for Saddam: death by hanging. No doubt for some, the happenings in Bulford were insignificant by comparison.

Private Gareth Aspinall had been a member of the multiple which had arrested Baha Mousa and the nine other detainees. He was also one of the soldiers assigned to guard the prisoners at various times during the forty-eight-hour ordeal that followed. The evidence he had given to investigators back in October 2003 and reconfirmed twelve months before the trial was crucial to the prosecution's case. Along with several other unit members, Aspinall was there to condemn the leading protagonist, Corporal Donald Payne. His statements were packed with detail about the abuse and those involved. The first had been made a couple of weeks after Baha Mousa's death and it had laid bare the systematic violence meted out to all the detainees. Corporal Payne was named as the chief protagonist, but others were implicated. A shadow of suspicion was cast over a whole range of responsible military personnel.

Julian Bevan QC must have expected Aspinall to reiterate his previously detailed statements. But another transformation had taken place. He was afflicted with a by now familiar memory loss, although it seemed to have bitten deeper for Gareth Aspinall.

Standing in the witness box, Aspinall calmly claimed he couldn't remember what happened, couldn't even remember what he'd said in his statements. The death of Baha Mousa, the treatment of all the other detainees, meant little to him. With automaton-like persistence his reply to Bevan's questions was the same: 'I can't remember.' Nothing could make him deviate from this serene forgetfulness.

Bevan became exasperated. He kept turning to the judge in apology and then in desperation. He couldn't extract the slightest cooperation. Aspinall had ceased to function as a viable witness of anything. Eventually, holding up his hands in frustration, Bevan had to sit down. He had never come across such obdurate memory loss before. He could only cling to the condemnatory impression that the original statements might have given to the military jurors. Now the defence attorneys had their chance.

Tim Owen led off as usual. He had watched in a state of amused astonishment whilst Bevan had struggled to obtain *any* evidence from the witness. It was a development that he couldn't have foreseen, not this extreme anyway. His cross-examination had been prepared to unpick the inconsistencies and inaccuracies in Aspinall's statements. His aim would have been to discredit the testimony against his client, but much of the work had now been done for him. Even so Owen needed more. His client had already pleaded guilty to inhuman treatment of civilians. He would be punished for that whatever happened. The charge he feared more was manslaughter. This remained on the indictment. And it was this that could send Payne to prison for years. Owen needed to weaken Aspinall's original statement further. The witness had to be tested and found so unreliable that the original statements would be dismissed as fundamentally flawed and fundamentally untrue.

As Bevan sat down with theatrical despair, Owen stood and looked at the witness for a moment. The time was 2.17 pm on 6 November 2006. It took him thirty seconds to slice open Aspinall's carefree amnesiac composure.

'Mr Aspinall, I don't actually have many questions for you,' Owen began, 'but let's try to make sense of your evidence. Of course, any question I ask you today will be met with the same answer, will it not: "I cannot remember"?'

'Yes.'

There was a suspicion of comic snorts from the watching benches.

'You seriously expect these gentlemen and everyone in court to believe that you genuinely cannot remember anything about this incident? You seriously expect us to believe that, do you?'

'Yes.'

'You took the oath and said you would tell the truth, the whole truth and nothing but the truth?'

'Yes.'

'Does that mean anything to you?'

It was a simple provocation, obvious, predictable, one of the stock questions barristers pose. Seldom does it produce a helpful reaction. Most witnesses are prepared and would say, 'Of course' or similar, sounding offended at the implication. The lawyer will then have to make a more pointed enquiry. Aspinall hadn't been coached. He was on his own. There was no advocate to watch out for him. He had lost all the protection that he could expect from the prosecution. In a sudden blaze he burst out: 'Don't you dare stand there giving me grief . . .'

What was left of his credibility cracked with the inappropriateness of the remark and the vehemence with which it emerged.

The judge calmly told Aspinall to answer the questions put to him.

After a few moments of silence, Owen began again.

'OK. You then proceeded, when you were asked further questions by Mr Bevan, to tell a pack of lies, did you not?'

Another provocation, a blanket accusation lacking specifics, a tactic easily rebutted. Aspinall was floundering. He didn't have a smooth riposte ready. His answer was instinctive.

'Are you calling me a liar?'

'Yes. And I suggest actually an embarrassing liar.'

Aspinall suddenly retreated into himself. He was wounded. Deeply. Years later that would be the insult he remembered most. Whether it was the accusation of lying that upset him or the suggestion that he had embarrassed himself in public he wouldn't say.

'Whatever,' he said, and adopted a hunched silence, seemingly suppressing the implicit violence of before. It didn't last.

'Mr Bevan asked you whether you had any memory of a man dying in custody while you were there or thereabouts and you told the president and members that you had no memory of that at all.'

'No.'

'That was an embarrassing lie, Mr Aspinall—'

'*You* are embarrassing,' Aspinall shouted, 'you're embarrassing, so sit down because you're getting up my nose.'

The courtroom stupor lifted. Everyone sat forward, eager to watch from their positions of safety as though drawn to an impending fight. The promise of impossible violence quickly faded.

'Have you had medical assistance about your memory difficulties?' Owen asked disingenuously.

Aspinall surprisingly remained still although he was clearly bristling. Perhaps he didn't pick up on the irony. He denied he needed any help. And then he stayed quiet.

Owen continued to stalk him. He brought out Aspinall's original statement and began to pick away at the details, looking to discredit every element of the witness' character even further. Aspinall's admission that his unit called themselves 'the Grim Reapers' was repeated. Before, it might have read like one of those nicknames which close-knit military units think up, childish, probably harmless, a minor act of bravado, little more. Like the names the troops gave each other: 'Doghead', 'Gorgeous', 'Asp'. Or the monikers adopted by playground bullies. Now there was menace lurking in the adopted epithet. Owen wanted to make Aspinall appear like a dangerous infant, a boy who couldn't be taken seriously when he accused others. He wanted anyone looking at Aspinall's youthfully chubby features, listening to him talk, reading his statements, to see and hear an undertow of violence and mendacity. By contrast, Donald Payne, a veteran of Northern Ireland, a soldier who had accepted responsibility for *some* of what had happened at Battle Group Main, could appear as a man honest enough to admit to misdeeds. But not so naïve as to become a patsy for others.

'He's accepted his responsibility for what he admits he did,' Owen said. Aspinall claimed he knew nothing about Payne's plea. He had only recently been flown back from Iraq for this trial and had been serving there for the previous six months. He didn't care what was happening back in Britain. He didn't care about the court martial.

Owen's next provocation wasn't subtle, but with this witness, subtlety had little purchase.

'You and your colleagues in the Grim Reapers were very fortunate not to be charged with any crime, I suggest, and you know that, do you not?'

'No.' And then, as though seeing through the stratagem, the penny dropping with clichéd perfection, 'Just hang on a minute, what're you trying to say to me?'

'I'm suggesting to you, Mr Aspinall, that you and the others in the Rodgers multiple are very lucky men.'

There: the intimation of criminal culpability. If Aspinall had been represented by his own counsel there would have been an objection. Just what was he being accused of? But the prosecution had been betrayed and there would be no intervention on his behalf. Aspinall responded as he must.

'Don't you dare stand there and accuse me of things. D'you know what I've been doing for the past six months, you stupid idiot?'

There was a flurry of voices.

Owen: 'I'm well aware you've been in Iraq—'

Aspinall: 'I've been back two weeks for me to come here and you give me grief—'

Owen: 'I'm asking you about events in 2003 which you know full well about, do you not?'

Aspinall: 'Whatever.'

With something of a delay, as though composing himself before intervening, the judge adopted the role of schoolmaster with a lawyer's practised disdain.

'I suggest, Mr Aspinall, that you behave.'

'I am behaving.'

'You do not call people a "stupid idiot". You listen to the questions and you answer them.'

'He's calling me a liar.'

'Do you understand?'

'He's calling me a liar.'

'Do not answer back. It will not do you any good.'

Aspinall assumed the demeanour of a recalcitrant child. 'I wanna break. I wanna break.'

'You are not going to have a break,' the judge told him.

And Owen resumed his questioning. 'Not long to go, Mr Aspinall,' he said rather too cheerily.

His next question was met by complete silence. Aspinall had his head down. He was lost in a game where he didn't know the rules and didn't even know what he was playing for. He had just returned from Iraq where, despite the danger, he had felt safe, safe with his mates, safe with his duties, safe in the environment controlled by order and discipline. But he knew nothing of this legal world. He didn't know how to act his part here.

'You're going to refuse to answer now, are you?' said the judge. Aspinall didn't have the fight for further confrontation. The pressure of all those faces behind their sleek surfaced tables, the quiet condemnation reverberating about him, the dark suits and braided uniforms, the reporters with open notebooks scratching his humiliation, all suppressed his indignant little revolt.

'No, I can't remember. I can't remember that incident . . . what went on. I've been through a lot since then. Personal problems at home and stuff like that. I can't remember what went on. A lot's happened since then.'

'Yes, all right,' Owen said, giving the impression of tiring of the verbal spearing that masqueraded as cross-examination. He had one final point to deliver. It was really the only point he had to make, a point that the judge already understood, that the panel of military jurors understood, that all the advocates understood. It is part of the unwritten rules that he had to restate it for the record. It was simply this: Aspinall and his pals in the unit had decided that Payne had to take the blame. He was the man they feared, the one who had orchestrated the abuse. If anyone was going to fall it had to be him. So they had invented a version of events, deflecting attention away from them and on to Payne. They had sworn that Payne had tried to organise a story immediately after the death had occurred. He'd approached them and told them what to say to the RMP when they arrived, trying to make the killing look like an unfortunate accident, the result of a failed escape attempt by Baha Mousa. In clubbing together to single out Payne's attempted fabrication they had pointed to the corporal's guilt without direct condemnation. That was the defence argument.

'Mr Aspinall, are you going to answer my questions now or not?'

'Questions for what – yes, I'll answer your questions. I'm being honest when I say I can't remember.'

'You gave in your statement – the first statement you gave on 10 October – would you look at page six? Do you see halfway down you are dealing with events after the death of Mr Mousa, yes?'

'Yes.'

'And then you say this, you put yourself in a little huddle of soldiers when Corporal Payne comes across and says "If anyone asks, we were trying to put his plasticuffs on and he banged his head or words to that effect," yes?'

'Yes.'

'That's what it says: "By the way Corporal Payne spoke and looked at us I took this to mean that that was what we were to say and nothing else." Yes?'

'Yes.'

'That's another little example, I suggest, of the Rodgers multiple lying to stitch up Don Payne, correct?'

'No.'

'You were going to put in the statement the suggestion that Don Payne was trying to get you all to give one account of what had happened, correct?'

'No.'

'And you, I suggest, Mr Aspinall, were never even present when Corporal Payne discussed the circumstances in which Mr Mousa had died, were you? Or perhaps you'll say you can't remember that either?'

'No. If it says that in the statement then that's what I said at the time. And it would've been truthful.'

'It was a lie, Mr Aspinall, like all of the evidence you've given today.'
Owen sat down. Aspinall was released.

13

P RIVATE ASPINALL WASN'T THE FIRST of the amnesiacs, or the last.
After him came Private Garry Reader. Twenty years old when he
was in Iraq, now twenty-three. He testified to the nicknames the
multiple gave to the detainees. 'Grandad', 'Bruise' ('because after some
time, bruises started to appear on his body'), 'Boy', 'Pisspants', 'Father',
'Fat bastard'. It was easier than using the detention numbers the prisoners
were allocated when registered in the logs.

As with all the soldiers who came to give evidence, Private Reader's
original statement from October 2003 had to be addressed first. This was
the one he had made to the SIB after he'd come forward to tell his version
of events. Before questioning got underway, he wanted to register a protest.
He said he had felt very pressured when he'd made his statement to the
SIB.

'The officer writing the statement put things down that I wouldn't have
worded like that,' he announced and intimated that the SIB officers who
recorded the statements from A Company had been putting words in the
mouths of the soldiers. What did this mean? Had he lied? Was it all a
fabrication, a concoction of an eager detective? No, but things hadn't
exactly happened as they were written down.

So what could Reader confirm now? Well, he saw nobody kicked or
punched. Yes, he was on stags for the night of the 14th and during the
15th but he didn't see anything to speak of. The detainees weren't allowed
to sleep, they had to march one or two of them up and down before questioning, that was about all.

Mr Bevan, conducting the examination in chief, encouraged Reader to
tell him about the night Baha Mousa died. Reader described seeing Mousa
outside the middle room of the detention centre, Corporal Payne and
Private Cooper grab him and take him back inside. He heard shouts and
went inside to see Baha Mousa lifeless on the floor. He described his
attempts at resuscitation, but he had little to say about how the detainee
came to collapse.

In the middle of a string of questions, Bevan tried a tactic of disorientation, what one might call a 'sudden lure'. He threw in a question completely unrelated to the death struggle.

'Ever heard of the choir before, or the chorus?'

Reader was taken off balance, but he replied that he hadn't heard this term before. Bevan's trick had failed.

When Reader was cross-examined he was a little more expansive. He agreed that when he saw Baha Mousa outside the middle room he looked as though he were escaping. He agreed with that interpretation, which was a crucial element of Corporal Payne's defence against the charge of manslaughter.

After Reader came Corporal John Douglas. He remembered the stress positions, but they were maintained with 'very minimal force', he said. The prisoners gave them no trouble really. A few slaps from some people to keep them in position, that was it; just a lot of shouting.

He didn't feel sorry for the Iraqis, he said. Why should he feel sorry that some were getting a couple of slaps? We had men killed out there; there was no love lost. He'd seen Baha Mousa being restrained by Corporal Payne the night the detainee died. He had seen Payne use a couple of slaps – definitely not a punch – the detainee fell and that was all.

Tim Owen, counsel for Corporal Donald Payne, had a different point of attack. Corporal Douglas hadn't been mentioned by anyone who had been in the detention centre at the time Baha Mousa died, nor as part of any guard in the preceding hours. No one could remember him being there at all. It was a mystery. Owen suggested Douglas had constructed a fantasy about his participation in the whole affair. Douglas couldn't remember the stag list and was confused about who else was in the detention centre at the time of Baha Mousa's last struggle. Owen kept repeating 'You weren't there, were you?' He portrayed him as a 'nowhere man, drifting in and out and no one notices'. Had he felt left out and wanted to be part of the 'incident' by claiming to have been in at the end? Why would he do such a thing?

Private (now Lance Corporal) Lee Graham was eighteen in 2003, now he was twenty-one. Like his A Company predecessors in the witness box, his memory was failing him. And before long he too wanted to volunteer information about his original statement.

'I felt that I was getting led a bit by SSgt Jay, the bloke who took the statement.'

It was a remarkably similar offering to Private Reader's. And now the

kicks mentioned in his original account had become 'digs', punches had
transformed into 'slaps'.

The judge suddenly called a halt to proceedings. He let the jury panel
leave the chamber and he released the witness for lunch. The advocates
were asked to remain for a private chat.

'I thought I ought to call a halt there,' the judge said, 'because I
thought I should give my assessment of the witness. It doesn't seem to me
this is a genuine loss of memory. He's reluctant to tell his tale.'

It was an obvious diagnosis. A disease had afflicted the witnesses from
Anzio Company: incurable amnesia. Had they come together and agreed
this approach? Or had someone got to them? Had they been coached to
say 'I can't remember' to everything of importance? Had someone
suggested that they claim the SIB officers pushed them to make exaggerated statements? The way in which two of them had 'dropped' into the
examination in chief a comment on the way in which their statements had
been drafted was too coincidental to be credible.

What could be done? There was no ready answer and no obvious remedy.
The judge had no option but to continue with the proceedings and allow
the prosecution to treat LCpl Graham as they had Aspinall, as a hostile
witness. But this changed little in truth. LCpl Graham was as obdurate as
his predecessors in the box. It was farcical.

Private Aaron Cooper, the soldier initially arrested as a suspect in the
murder of Baha Mousa, was not quite so badly afflicted. He could recall
the 'choir' but as far as he was concerned only 'moderate force' had been
used, nothing extreme. He said Corporal Payne had prodded the detainees
with the barrel of a gun. And there was a slap here, a slap there, just to
keep the prisoners in position. He admitted some visitors used greater
force against the Iraqis, striking them with their fists, but none of them
were from his multiple. Later, on the Monday night, he had heard Corporal
Payne shout for help. Cooper said he hurried to the middle room of the
detention centre and saw Payne struggling with Baha Mousa. Cooper had
lent a hand, trying to get plasticuffs on to the prisoner. He thought Payne
had kicked the man in the ribs and hit his head against a wall.

Tim Owen, with his usual robust character aflame, began his cross-
examination by claiming Cooper was a liar, 'a practised and confident liar'.
Why? Because he was desperate to avoid being implicated in any charge.

Cooper denied the allegation. He said he was only trying to protect Corporal Payne when he had lied to the SIB soon after the death. But he agreed with Owen that he thought Mousa had been trying to escape when he'd helped restrain him. He did accept that he lied in his interview; he hadn't told the SIB that Corporal Payne had hit the detainee's head against a wall, he hadn't said Payne had also punched one of the detainees, as he claimed in later interviews. Those were all lies.

Similar admissions of lying throughout his interview with the SIB all those years ago were extracted. Cooper became confused over whether Payne used a barrel of a gun or his index finger during the 'choir'. His previous statements never mentioned a gun before. Owen concluded with a rhetorical flourish, admonishing the witness for being an irrepressible liar only intent on protecting himself from any charge connected with the events at Battle Group Main.

Cooper wasn't the last to suffer from amnesia. Some were worse than others.

Private (now Lance Corporal) Slicker hadn't been part of Anzio Company. He was in the quartermaster's stores in Battle Group Main. His memory was so poor he couldn't even remember the months he went to Iraq. Deployed to a war zone, an intense environment that all agreed bore no equal, and still all details seemed to have been erased.

Could he remember the screaming? No.

Could he remember going to the detention block? Just about.

Could he remember seeing Iraqis inside? Sort of.

Could he remember how many Iraqis there were? No.

Could he remember whether they had plasticuffs on? No.

Could he remember what was making them shout and scream? Punches.

Could he remember who was doing the punching? No.

Could he remember who was in the detention block? Lieutenant Rodgers and his multiple.

Could he remember their names? No.

Could he remember seeing Rodgers punch anyone? Yes.

Could he remember when he found out a man had died? No.

Could he remember seeing something at the generators behind the main buildings on the base? Yes, he did. He remembered seeing a detainee kneeling beside one of the generators with a sandbag over his head. He couldn't remember who was guarding him.

But could he remember being disciplined for hitting one of the detainees?
Yes.

Could he remember how he was punished? Fined £600 by his
commanding officer for battery of an unidentified Iraqi male.

Could he remember what he did to earn this punishment? He 'tapped'
him with his foot.

Could he remember why he did this? Truthfully? He was pissed off.
Pissed off because there's no justice. Nine RMPs murdered, Iraqis shooting
everyone, getting away with murder 'but us, one person dead and everyone
gets blamed for it. So where's the justice in that?'

14

THE COLLECTIVE FORGETTING AND GENERAL failure to provide
witnesses of sufficient candour and credibility had weakened the
prosecution's case against those accused of violence. But if the direct
accusations of assault and manslaughter and inhuman treatment had been
undermined, the charges of institutional failure had not. Even though
flawed, the testimony, or rather non-testimony in many cases, had revealed
how the violence against the detainees had been casual, callous, and visible.
That no one had protested about the treatment of these prisoners at the
time, and only a few had come forward after the event to register complaint,
suggested a culture of abuse entrenched in the battalion. This was the
second major claim of the prosecution.

Who, then, should take responsibility for this culture? The prosecution
had Colonel Mendonça, Major Peebles and WO Davies in the frame; the
colonel as commanding officer, Major Peebles as the officer directly responsible for the regime of conditioning, and WO Davies as the trained tactical
questioner. Surely, these men couldn't escape censure? They had been in
command of a whole unit incapable of understanding right from wrong
at a most basic level, or so it seemed. This was the prosecution's contention
and the reason why they had been chosen to stand trial.

There were difficulties here too. Could anyone say with clarity whether
there existed a recognised and proper system of treatment for all detainees,
one which the accused should have understood and which protected the
welfare of any prisoner?

Major Ed Fenton was called to provide part of the institutional story.

He was the author of the 'chaps' email message immediately after Brigade had found out about the death of Baha Mousa. Fenton had sought advice and information, trying to bolt the stable door, as it were. He admitted now that during his post-death investigations he couldn't discover who was supposed to have been responsible for prisoner welfare. Nevertheless he was adamant that interrogation was an essential fight against the insurgency. It couldn't excuse the treatment endured by these detainees, but it perhaps made it understandable. The pressure had been on for information, he said. There were complaints about delay, about the 'wrong' men being sent on to Camp Bucca, innocents who had nothing to say and posed no threat and whose transfer to the internment camp had been a waste of everyone's time.

In Fenton's cross-examination, Lord Thomas for Major Peebles stressed that his client should have been able to rely on the good training of others under his command, that he should have been confident that the provost staff who looked after the detention facility, the tactical questioners who came over from Brigade, would all be adequately, if not expertly, trained in handling prisoners properly. Major Fenton didn't demur.

Then Major Royce appeared.

Major Royce had been in Iraq as 1QLR's BGIRO from June till August 2003. He was Major Peebles' immediate predecessor. He'd had nothing to do with the treatment of the detainees, had never seen Baha Mousa and those other men, and had been back in the UK when they had suffered their torture. But he was to be instrumental in determining whether the case against Mendonça, Peebles and Davies was to succeed.

Major Royce had given a statement to the SIB several months after Baha Mousa's death, but it hadn't been favourable for the prosecution's arguments. In fact, it was fundamentally damaging. It had been disclosed to the defence, as it had to be, but presented by the prosecution as flawed testimony. Julian Bevan therefore applied to treat Major Royce as a hostile witness. It was the only means by which Royce's destructive evidence could be challenged by the prosecution.

In an unusual move, though, the witness was called by the judge in order to be cross-examined by *all* the advocates, Mr Bevan included. And it was Bevan who had the first go.

What could Major Royce say? He had been 1QLR's Battle Group internment review officer when the battalion had first arrived in Iraq.

Like Peebles he had been given no training for this role. He'd had to make things up as he went along. His handover from the Black Watch regiment had been minimal, but he'd seen detainees hooded by their men, something which worried him. He said he'd sought guidance from two of the officers at Brigade, Major Robinson, Brigade intelligence officer, and Major Russell Clifton in the legal team. He'd wanted to know whether the practice of hooding was acceptable. They told him it was 'OK', or so he claimed.

On the meaning of 'shock of capture' which was supposed to be applied to new prisoners Royce said he spoke to Brigade about this too. He'd been told that it was to be used for those detainees who were to be tactically questioned, that it could involve hooding and stress positions 'in order to continue the disorientation, unease, of the prisoner'. Royce was clear: Brigade legal told him this was acceptable. Bevan couldn't quite believe it.

'Are you telling this court that you had cleared that process with Major Robinson and Major Clifton?'

'Yes, that's correct.'

Armed with this advice, he said, he'd instituted the regime of hooding and conditioning, stress positions, plasticuffs, everything the court had heard in relation to the detainees and Baha Mousa's death, that had been lamented as systemically illegal, unacceptable behaviour, provoking the prosecution of Colonel Mendonça, Major Peebles and WO Davies, everything bar the kicks and punches that is. He had told the guards what to do, Royce said, given them instructions that the prisoners should be forced to maintain stress positions. And how was this to be done? Well, they would have to manhandle them somehow. That was inevitable. It didn't mean they had to abuse them. They were maintaining the 'conditioning'. It was a very fine line for the young soldiers involved.

About seventy detainees had been processed by the time he had completed his tour, Royce said. He was adamant that he'd cleared the whole process with Brigade. Why had he done this? He'd been worried. That was why he'd asked. And, he said, just to be sure everything was above board, he'd told Colonel Mendonça that the approach had been given clearance. When Major Peebles had arrived in theatre to take over the BGIRO role, Major Royce had briefed him fully on what was to be done. Everything about the hoods, the stress positions, the shouting, the conditioning, the shock of capture, all of it relayed so that Peebles could follow standard procedure. Indeed, they had been lucky, Royce said, to have had a couple of detainees arrive at Battle Group Main during the handover to Peebles to show him properly how it all worked. These prisoners were placed in stress positions ready for questioning. It was all very fortuitous. Major Peebles had been given first-hand experience of what to do.

In as subtle a way as possible Bevan suggested that the witness was mistaken, at least about the stress positions. Major Robinson hadn't given clearance, had he? Major Clifton hadn't given clearance either. Royce remained steadfast. He was certain about the advice he'd sought and received. Reliant upon this, he'd instituted the system that 1QLR applied. What reason would he have for lying? To protect himself? But he hadn't been targeted for prosecution, as far as anyone knew. There was no threat of disciplinary action, not now.

Without any obvious motive for deceit, Royce's evidence was poisonous for the prosecution. The charges against Colonel Mendonça, Major Peebles and WO Davies presupposed that they had either known the regime of ill-treatment used against the detainees was wrong or hadn't cared. This would only be convincing if they had been acting without regard for army rules. Major Royce asserted that that supposition was flawed; the command *had* looked at the methods used and *had* approved them. Even if declared later to be 'illegal' with hindsight and many legal minds poring over the details, the charge of neglect couldn't stick if Brigade had sanctioned the regime imposed on the detainees. If Royce was to be believed, it was Brigade who should bear the responsibility. Their sanctioning of stress positions and hooding was specific.

All of the defence lawyers were friendly to Major Royce. They lapped up his evidence and each in turn found some succour for their client. He confirmed that all the hooding and stressing and conditioning were in accordance with agreed and fully sanctioned protocol. It didn't mean that the prisoners *should* have been treated in this way. But it did mean that the individuals on trial should not bear responsibility for doing what had been ordered.

Major Russell Clifton followed immediately after Major Royce. He'd been based at Brigade headquarters at Basra Palace during his tour in Iraq as the senior member of the legal team. The first question Mr Bevan asked was about hooding: what was Brigade's policy?

Clifton believed that no hooding should have taken place. Did he know Royce? Yes. Did he recall any conversation with Royce about hooding? No. If he *had* had a conversation with him about hooding, he would have said it was forbidden. But he would also have said that stress positions could be used before tactical questioning. Did he have a conversation with Major Royce about that? He might have done, he said.

It was now Mr Owen's turn. He wished to challenge the dichotomy Clifton had presented: hooding, definitely not acceptable; stress positions, could be OK as part of conditioning. But, Owen said, Major Sian Ellis-Davies, the lawyer at Division level, the woman officer who had advised internment officers about any legal issues arising, the woman who had been one of the first to learn about the fate of the detainees when they had been delivered to Camp Bucca all those years ago, had advised the court that stress positions were undeniably *illegal*. There was no way she would have authorised them, she would have 'gone mental' she had told the court. How was it then, Owen asked, that Major Clifton at Brigade was giving a wholly contradictory response?

Clifton couldn't say. He was at a loss. No one seemed to have a clue what the rules were or should be.

Bevan called Major Robinson, the intelligence officer at Brigade and from whom Major Royce had said he'd received clearance about the methods of detainee treatment. Major Robinson was quickly taken to the hooding and stress positions issue. He too was ambivalent. Unlike Clifton, though, he said hooding was not the subject of specific policy. They all knew it wasn't allowed for sensory deprivation, he said, but for a 'brief period' it would be all right, for security reasons and such like. Stress positions, however, were definitely not acceptable, he said. Sleep deprivation? Perhaps as a consequence of the shock of capture or if prisoners were being moved from place to place, but otherwise it was outlawed. Why was this so? Major Robinson said he knew about the Northern Ireland ruling and he knew the techniques used then were banned.

Under cross-examination, Robinson was asked by Mr Owen to explain why Major Clifton, the brigade legal officer, a man with whom Robinson worked on a daily basis in Iraq, had given the court the exact opposite opinion to him, that stress positions would be acceptable in certain circumstances. How could they be so diametrically opposed?

'I would put that down to Major Clifton's inexperience in an infantry environment. He purely has a legal basis – that was his forte.'

'If he *had* a forte,' Mr Owen replied rather acidly.

Lord Thomas on behalf of Major Peebles turned on Major Robinson. He wanted to know how Robinson came to be aware of the outlawing of the five techniques. It was on a training course, Robinson explained. In 1998 he'd been sent to Herefordshire where he was sent out into the countryside, dressed only in 1940s battledress, left to fend for himself without any rations and with the instructions to avoid capture. A team playing the enemy searched for him. And, of course, they had found him and then subjected him to the hooding, the stress positions, interrogation, harsh questioning. They were trying to break him, to make him say more than his name and rank. It was after that experience that he was told that the techniques used against him had been against the law. They weren't supposed to form part of any detainee treatment, but it was thought he needed to know what it was like just in case he was captured by a less scrupulous enemy.

Lord Thomas picked up a book from the desk in front of him. It was the journalist Robert Fox's account of the Falklands War. It was handed to Major Robinson and he was asked to turn to a photograph on page 130. The caption was 'A lieutenant commander of the Argentinian Marines having been captured by 40 Commando at San Carlos'. It showed a man in a jumpsuit, his hands tied or pinioned behind his back, standing between two British soldiers. His head was covered in a hessian sack of some kind, tightly bound around his neck.

Did the picture shock Robinson? Lord Thomas asked.

No.

Prior to the death of Baha Mousa would he have found hooding like this acceptable?

Yes, but only for limited periods. And he didn't like the look of the Argentine's hands pinned behind his back.

But this was taken in 1982, wasn't it?

Of course.

So it was ten years after the government directive outlawing the five techniques used in Northern Ireland internment?

Yes. But that related to treatment *prior* to interrogation.

So hooding wasn't outlawed completely.

No.

Another photograph from another book on the Falklands was produced. Another photograph of an Argentine prisoner with a hood over his head. The caption underneath said that the man was on his way to be interrogated. What did the major make of this?

Same response as before. Hooding might be necessary for security reasons.

Returning to Iraq twenty years later, what about sleep deprivation? Would he have been shocked to know that Sergeant Smulski, one of the two tactical questioners sent to Battle Group Main and under Major Robinson's command, had interrogated the detainees throughout the night and had ordered them to be kept awake?

Yes.

Had he ever gone to see what the tactical questioners got up to?

No.

Did he know what techniques these tactical questioners had been trained in?

No. That wasn't his job. Although he commanded Smulski and Davies, their roles as tactical questioners were outside his remit. They were trained TQers and were called on by battalions when needed.

Hadn't he been curious about what they did?

No. He was busy and it hadn't occurred to him to ask. He received information from the TQers and that was enough.

Major Robinson was asked to think again. It was put to him that either he had no grip on the intelligence cell personnel he commanded, knowing little about what they were doing, or he *did* know, but now chose to forget. Robinson refuted both accusations.

The conflict now between the evidence from Royce, Clifton and Robinson, all majors, all senior officers, all giving different accounts and different interpretations of what was right and what was wrong, what had been said between them, was a portrait of utter confusion. There appeared to be no collective appreciation of how any detainee should have been treated. If that was true, how could Colonel Mendonça or Major Peebles or WO Davies or any of the other troops ordered to apply conditioning to the detainees be held responsible? How could any of them know what was right and wrong if those in charge of policy couldn't agree?

15

THE CHRISTMAS RECESS WAS DRAWING closer. There would be a struggle to complete the prosecution case before the prolonged adjournment. It would fail and the trial would continue deep into the New Year.

Captain (now Major) Gareth Seeds was called to tell his story about entering the detention facility sometime after the death of Baha Mousa and seeing the prisoners in a poor state. He presented his testimony in a measured fashion, nothing hyperbolic, nothing beyond a simple description of what he'd done and what he'd seen. He recalled something else as well. On the Monday he'd witnessed a detainee kneeling in the sand not far from a hot and noisy generator in the yard at the back of the accommodation block. The man was hooded and his hands were cuffed behind his back. It must have looked like one of those pictures of a battlefield execution. But it seemed perfectly unobjectionable to Seeds at the time. He'd assumed it was a procedure handed down when the battalion had taken over Battle Group Main.

He did have some words to say about Colonel Mendonça, though. It developed into quite a tribute.

'Without Colonel Jorge,' Major Seeds said, 'the battalion would probably have failed. Y'know, he was a shining example. We all tried to live up to his abilities and, y'know, we all fail because we're lesser men. I'm very honoured to have been his operations officer throughout the tour. He was the right man for the right job and without him I think we would've lost more than the one we did.'

The case against Colonel Mendonça had been feeling hollow for a number of days now. Major Seeds' testament reflected a persistent theme portrayed by the officers called to give evidence. No one flecked their testimony with any suggestion of criticism directed at the commanding officer of 1QLR. Perhaps Major Seeds' homage was the most effusive, but it didn't discord with anything others had said.

The military panel, some of whom were out-ranked by Mendonça, must have wondered about the charges again. Some may have seen in him a model of the modern professional British soldier: awarded the DSO and promoted to full colonel as a direct result of his service in Iraq. How could this have been compatible with the charge of negligence?

The eulogising continued with the appearance of Major Englefield, the overall commander of Anzio Company. When asked his opinion of the colonel, he said 'He had a very firm grip on his officers and his soldiers . . . he was a very strong and disciplined leader. We knew where we stood . . . but he was approachable too . . . he was on the ground a helluva lot . . . only the greatest professional standards would be tolerated . . .'

16

S UCH WAS THE ERRATIC EBB and flow of proceedings that for a moment attention shifted to other defendants. Witnesses were brought forward as they became available. It made for a confusing pattern, but there was no choice given the administrative difficulty of bringing to court so many military personnel from across the globally stretched operations of the army.

The crucial moment for LCpl Crowcroft, Kingsman Fallon and Sergeant Stacey had arrived, though. It was 28 November 2006. Private Jonathan Lee, the whistleblower from the Hollender multiple of Anzio Company, was to give his evidence. He was the one who had said he saw all three accused assault the detainees on the first day of their arrest.

Over the course of the next two days, Private Lee was interrogated like a slowly roasting pig.

He began adamant about what he had seen, kicks administered by Lieutenant Rodgers to the Iraqis lying on the floor when arrested in the Hotel al-Haitham, 'like kicking a football', then in the detention centre back at base, Corporal Payne, whom he knew, came and went without perpetrating any violence, and Stacey, Fallon and Crowcroft were left behind to look after the prisoners. He did remember that during his short time in the facility about twenty other people of differing ranks had come in to take a look, but it was Crowcroft, Fallon and Stacey whom the court was most interested in. What were these soldiers doing?

'They were shouting and swearing . . . Crowcroft and Fallon were going round punching and kicking the Iraqis . . . every time a person went to ground, they went over to them, pulled them back to their feet and then waited for the next one to fall . . . Corporal Stacey was laughing . . . when I said the person to my left was doing quite well in the stress position, he kicked his feet until he fell to the floor . . . he kicked the bottom of his feet . . . he stamped on him and punched him and dragged him back to his feet.'

After these moments of cruelty, Sergeant Stacey had taken Lee away from the facility back to their barracks, leaving Crowcroft and Fallon behind. Later on in the day Lee had seen the two junior soldiers again, back at Camp Stephen, the sister base of Battle Group Main, and they told him that they had been left to beat the prisoners up. The two soldiers 'had cuts and bruises all over their hands and their shins and their feet', Lee said. 'They were complaining about it.'

Perhaps Mr Bevan, who was undertaking the examination in chief, could feel the air of scepticism surrounding Private Lee's testimony. Lee was a young lad after all, only twenty-three years old, and he had the appearance of an adolescent still, maybe not wholly convincing to the panel. Bevan knew that Crowcroft's and Fallon's barristers would question why it was that Lee had felt the urge to inform on his colleagues. But Bevan didn't want to unravel the knot of motivation. He simply asked when Private Lee had first told someone his story.

It was when the regiment was serving in Cyprus, March 2004, Lee said. He had gone to see the doctor about a spell of depression and told him what he'd seen in Iraq. The doctor had put him in touch with the padre, the padre had sent him to see the RMP, and the RMP had contacted the Special Investigations Bureau. It was a little convoluted and certainly gave no explanation of why he had waited until that moment to say anything.

Mr Ferguson, LCpl Crowcroft's barrister, had hardly finished standing up before he began his assault on the witness. Private Lee had made all this up, hadn't he? His behaviour had been aggravating, hadn't it? He'd been moved from company to company because the other soldiers couldn't stand him, wasn't that right? He couldn't make it as an infantry soldier, struggled from the beginning, went AWOL for over a year between March 2002 and March 2003, for which he was found guilty at court martial, correct?

The last allegation was admitted by Lee. He had no choice as the record was plain. What had he been doing all that time when he had been on the run from the army? Ferguson asked.

'I was with my girlfriend sir.'

'That's hardly a full-time job, is it?' Ferguson said. 'Were you drawing unemployment benefit?'

'No, sir.'

'How were you managing financially?'

'I had savings, sir.'

'Savings? From where?'

'I had about four grand saved up from the army.'

What had he been up to for a whole year?

'I was seeing friends . . . out, sir.'

'Out?'

'Aye, sir.'

Why had he absconded in the first place? Private Lee said he'd been bullied. When he had handed himself in eventually, he'd been given fifty-six days' detention as punishment. That was when he was sent to the Colchester glasshouse, where he first came across one Corporal Donald Payne. It was a small world in 1QLR.

Ferguson continued to dissect Private Lee's character, his personality and its various failings, exposing them to the court and the press assembled about the chamber. It was ritualistic in its brutality. But he wasn't concerned with the wisdom in sending such a flawed soldier to a war zone. He wanted to press on with his prepared argument that everything Lee had said was suspect. He 'constantly shirked his duties', he 'frequently pretended to faint', he 'fell asleep on duty' were the accusations Ferguson threw at him. And, pointedly, he'd had an altercation with Crowcroft, one of the men he was now accusing of assaulting the detainees. Ferguson suggested Lee had refused to relieve Crowcroft from guard duty, Crowcroft had chased Lee down some stairs, Lee had dived into a guardroom, grabbed a rifle, cocked it, pointed it at Crowcroft, before another soldier had snatched it off him. And then Lee had shouted he would grass up Crowcroft. Did Lee remember this now?

No.

'I suggest that the account you've given about what took place on 14 September 2003 is a concoction of lies. You were never in the detention centre . . . that's something you made up . . . you're a fantasist,' Ferguson said.

'No, sir,' Lee replied.

Ferguson pressed him with further allegations. All Lee had done in this chamber was tell lie after lie. The accusations speared through the young man who acknowledged he had a troubled history. There seemed little left that was credible of the man whose original statement in March 2004, all those years before, had been instrumental in bringing three men in front of Britain's first official war-crimes court martial. If eyewitness testimony relied upon the quality of integrity then Private Lee had tested the connection to its limits, or at least, that was the impression Richard Ferguson QC wished to plant.

The other barristers acting for Kingsman Fallon and Sergeant Stacey took precisely the same approach. Lee was a 'habitual and congenital liar', Geoffrey Cox QC proclaimed. Everything had been made up to suit Lee's purposes. He had used the whistle-blowing as an excuse to return to Britain from Cyprus, to go on long-term sick, and eventually to leave 1QLR and the army.

Jonathan Lee spent a day and a half rebutting allegation after allegation. His fortitude before the prolonged assault was remarkable, a tour de force of resistance to the pressure applied by some of the most senior criminal law barristers working the courts today. Was it convincing? The jury panel consisted of senior army officers. What would they make of a young man who by his own admission wasn't cut out for army life, who was a failure, who went absent without leave almost as soon as he joined the army, who stayed on the run for over a year, who served time in an army gaol for his absence, who was released only to be sent immediately to serve in the cauldron of Basra? What would these professional soldiers, with the creed of the army deeply imprinted in their imaginations, see in ex-Private Lee? Would they consider him a liar? About everything, everything he claimed to have seen at the detention centre?

Lance Corporal Gareth Hill was in the same multiple as Private Lee. He was next into the witness box and he, like Lee, could remember when LCpl Crowcroft and Kingsman Fallon returned to base after they had finished their guard duty at Battle Group Main. They were speaking about 'the choir', he said, about people being hit in the stomach to produce differently pitched noises. But Crowcroft and Fallon weren't boasting, Hill said. They were just telling everyone what had gone on. When cross-examined, with little of the sustained assault on his integrity as Lee had received, Hill admitted he may have been mistaken about what he'd heard.

Perhaps with all the talk about the choir he was misinterpreting what had been said. Perhaps what he had actually heard was people's description of Corporal Payne's little joke, not boastful admissions of brutality. Might that be right?

Yes, it might, he agreed.

17

SHREWD COURTROOM LAWYERS KNOW WHEN a case is beginning to slide. They can smell it. In every break in proceedings they evaluate the notches of success and setbacks and think about the impact of a particular question or a particular answer. A mental reckoner registers the changing shape of a trial and gauges the likely result. Rarely is there the flash of evidence that shatters or makes a case in one instant. All those films and books reliant upon the moment of revelation represent the exception, not the norm. That doesn't stop the lawyer from searching for that moment. It's the moment that's savoured if achieved (rarely) and lamented if thought to have already passed (frequently). But the norm is more prosaic. Shifting perceptions of success or failure emanate from an interpretation of all those words and bodily signals that it is hoped convey messages to the jury.

What might Julian Bevan QC have thought as he concluded the presentation of witness testimony for the prosecution? He was perhaps too old now to worry. The thickened skin of the advocate had to allow for the wrong result and the judicial reprimand. It was a fundamental component of the good lawyer's qualities. Win or lose, his reputation should stay intact. Although in this case there had been such a degradation of witness evidence that accusations about poor preparation might have acquired some purchase in the Inns of Court. Later, for certain, there would be no mention of his contribution to this particular trial on his résumé on his chambers' website. The list of legal achievements were impressive: the Guildford Four, the Dennis Nilsen murders, the Pimlico killing of Lady Cross in the basement of an antique shop, the Lester Piggott tax fraud, the Kenneth Noye case, the 'Bakewell Tart' murder appeal, the Jubilee Line fraud trial. The Baha Mousa killing court martial wouldn't appear among these achievements. But he was still lauded as a 'vastly accomplished . . . renowned big hitter' in the website quotes from one of those annual guides to the legal profession.

Back at Bulford Court Martial Centre, Julian Bevan QC must already have sensed the exhaustion of the prosecution case. None of the defendants had been seriously prejudiced by the evidence he'd presented. It was a relief that Donald Payne had already pleaded guilty to inhuman treatment. Even that charge would have been under threat given the patchy nature of testimony heard so far. If conviction requires a belief of guilt that suffers no doubt, then too many uncertainties and suspicions and reservations had been introduced to be confident that the charges would be found proven.

Of course, the prosecution couldn't give up. There was still the chance that the military panel would be sufficiently horrified to believe that these men were guilty as indicted. There was still a case to answer, so Bevan must have hoped. Despite all the amnesia and contradictions, despite the confusion and chaos, despite the lack of understanding within the army about its obligations, its training, its systems, despite all the failings of personnel and institution, there was still one undeniable set of facts: Baha Mousa had been killed in custody and nine other detainees had been subjected to brutal treatment. Wouldn't that be enough?

Before Christmas 2006, the prosecution strove to introduce more witnesses who could contribute to the air of outrage. But when the junior medics at Battle Group Main were heard, they could add little to the stories of abuse. Their testimony confirmed that at 3 pm on the first day of detention an examination of the detainees hadn't revealed any injuries. That was Private Winstanley's evidence, evidence which helped Crowcroft and Fallon.

A number of army lawyers appeared next, anxious to reiterate the unlawful nature of hooding and stress positions. Chief amongst these was Lieutenant Colonel Nicholas Mercer. His story, though, was a little more complex, a little more revealing. He was replete with tales of ill-treatment of prisoners by British forces in the early days of the war. He had been flown into Camp Bucca in March 2003, soon after the invasion, accompanying some senior officers in their review of PoW facilities. He told the court that during his visit he saw about forty Iraqis 'kneeling in the sand, cuffed behind their backs, in the sun with bags over their heads'. He was 'surprised' he said. He thought this breached the Geneva Conventions and he brought the matter to the attention of his superiors. It was only the matter of hooding which prompted discussion. Mercer told the court that he'd had to fight to be heard even on this issue, that what he'd seen was unlawful. Others in the upper echelons of the army legal fraternity disagreed with his analysis, he said. He was told that the hooding he'd witnessed was 'British Army doctrine'. Hooding was fine, was the official position. It was legitimate to cover a prisoner's head to prevent him seeing the inside of an army facility, to stop him conferring with fellow prisoners.

Mercer said the International Committee of the Red Cross became involved when they had filed a complaint to the British authorities also criticising the practice of hooding for long periods. There had been a meeting, Mercer said.* The problem was resolved, according to Mercer, by an order from the chief of staff that hooding was no longer to be practised. This was in the middle of 2003, several months *before* Baha Mousa and the other detainees from Operation Salerno were arrested. But Lt Col Mercer told the court that the whole preparation for taking and dealing with prisoners was risible. There was confusion from top to bottom about what was legal and what wasn't. He had been in serious arguments with his colleagues about interrogation methods. Eventually he had left Iraq with his division in the summer, before 1QLR were deployed. And with him, he surmised, went the determined opposition to hooding and stress positions and questioning by capturing units.

Mercer's testimony succeeded in only one thing for the prosecution. It reinforced the perception that the army's understanding of its legal duties towards prisoners was a farrago. There had been less a 'corporate loss of memory', more a collective inability to distinguish between right and wrong. Major Royce's account remained believable. For the accused officers it was an absolution born out of moral disorder.

18

A S IF TO HEIGHTEN THE sense of a moral void and the folly in prosecuting a few selected members of 1QLR, the battalion padre was called to the stand.

Father Peter Madden was new to the unit in 2003. He had left his country parish shortly before the invasion of Iraq to join the army. His bishop had had to give permission. The padre received basic training and

---

* Lt Col Mercer wasn't able to reveal the full story about this meeting nor the order he received *not* to mention what he had seen in Camp Bucca. This only came to light in the public inquiry which followed some years after.

then left for Basra. Madden remembered that he used to visit the Portaloos outside the detention facility. Often there would be prisoners there. He would speak to the soldiers on guard duty, sometimes entering the facility. He couldn't remember seeing anyone in 'ski positions' though. A bit of squatting, he said, but you'd see so many people squatting in the streets of Basra, wouldn't you? No, he couldn't remember any hoods. Well, except for one occasion when he'd seen a prisoner being led along outside head-quarters. That man had been hooded, some kind of sack over his head. But that was for security reasons, wasn't it? he said. And he'd heard some occasional raised voices coming from the detention building. He would be curious and go and have a look, but there had been nothing untoward as far as he could see. He couldn't remember much else.

When cross-examined by Colonel Mendonça's barrister, the padre's memory glowed a little more warmly. The colonel was 'able and driven. Very on top of things. Very keen that everything be done properly. Scrupulously fair . . . it was a good regiment, a happy regiment.' He was a little aggrieved that the colonel hadn't invited him to accompany him on his visits to Baha Mousa's father after the son's death. He could have been of some use, he thought. But he was excluded, no doubt for good reasons. It wasn't for him to question that decision or any other.

That was the end of his evidence. Why had he been called? His earlier statement to the SIB during their trawl through those at Battle Group Main had suggested he might have witnessed some ill-treatment, at least confirming the commonplace hooding and stress positions. Perhaps it was hoped that he would lend moral weight to the prosecution's case. In truth he said nothing of value other than for the colonel's defence.

Should the padre have been quizzed more strenuously? In the context of a normal British criminal trial, which this case was apart from the mili-tary embellishments, the answer had to be 'no'. Father Madden's responsi-bilities for safeguarding the welfare of prisoners within his unit weren't under consideration here. Only the men accused had to answer questions of blameworthiness and conscience. All the others might have been asked why they hadn't intervened or why they hadn't noticed ill-treatment or whether or not they even cared what happened around them. The padre, as with many others acting the part of witness only, could avoid such probing with ease. And Father Madden adopted a character of gentle innocence that matched common perceptions of men of the cloth, even those who wear khaki. He was willing to concede that maybe he should have done more for the prisoners he knew were on the base. Maybe he should have attended to their religious needs, at least by ensuring that they were given space and time to pray. Maybe he should have sought out and entered the detention centre and checked on the welfare of the men detained. The padre could afford to be generous on reflection, but there were no consequences for his mild admissions. At this court martial he was merely the witness, and given that he saw little according to his account, his part was swiftly fulfilled. If it had been an inquiry, an inquisition perhaps, his moral neutrality would have been in greater jeopardy. Then he might have been asked whether he should have interfered. Although any intervention would have probably been rejected: when it came to operational matters the unit chaplain had no right to overrule command decisions. He might censure, offer his advice, but that was the extent of his role.

Father Madden hadn't even gone that far.

19

AIMLESSNESS NOW CREPT INTO THE procession of witnesses, as though filling the court's time with a succession of morally virtuous army officers would substantiate the prosecution's case. SSgt Sherrie Cooper, the first SIB officer on the scene after Baha Mousa's death, was called. Her testimony was ended abruptly when the defence barristers realised that she wasn't the Cooper they were expecting. She had travelled all the way to Bulford to speak about nothing for no more than three or four minutes. There were no questions about her investigation, no searching enquiry as to why she hadn't sealed off the crime scene, why it had taken so long for a crime even to have been identified, why possible suspects were not pursued, why the investigation had been subdued in those early hours and days.

Then the regimental doctor, Derek Keilloh, was called. He appeared on 11 December, although the weather was more autumnal than festive. Rain, gale-force winds and mild air from the south gave the days an unhappy sense of injustice amidst the fake holly and tinsel. He told the court that he had now resigned his commission and was practising as a GP in Northallerton. He didn't know of any procedures for the conditioning of prisoners, not at the time. And he denied going anywhere near the detention facility until he had been called to attend to one of the prisoners who had reportedly collapsed. That was Baha Mousa. He tried all he could to resuscitate him, he said, there at the facility and back at the Regimental Aid Post. He didn't see any injuries on the body except for a bloody nose. And no, he didn't make enquiries as to how that small injury had happened. He admitted that a little later on the night of Baha Mousa's death he had been asked to examine another of the detainees. The prisoner brought to him had complained of being beaten. Dr Keilloh said he gave him a painkiller. He hadn't asked how the injury had come about. He hadn't questioned the guards. An hour later another detainee had been brought to him. This man had made a similar complaint. Again the doctor had made no enquiries.

The prosecution barrister, Mr Barnard, filling in for Julian Bevan, couldn't avoid injecting a note of disbelief into his questioning of the doctor. Had he really not thought to find out how the detainees appearing in the RAP had been injured? This line of questioning attracted the objection of the defence attorneys. The prosecution were acting as though they disbelieved the evidence of *their own* witness, as though they thought the doctor hadn't discharged his duties, which, of course, could be treated as an offence similar to that for which Colonel Mendonça and Major Peebles stood charged. The judge admonished Mr Barnard.

Time then for the defence advocates to quiz the doctor. Only Julian Knowles for Corporal Payne had something to test and in doing so he showed he also doubted the doctor's testimony. He suggested that Keilloh *had* been inside the detention facility before he was called to resuscitate the stricken Baha Mousa. Dr Keilloh denied the allegation. Payne hadn't asked him to look at one of the detainees who kept falling over, he said.

That was the extent of the probing. Some of the defence barristers had seen the jury panel register expressions of concern, maybe even disbelief at what the doctor had said. Had he really not seen fit to investigate the allegations of the two detainees seen after Baha Mousa's death? Had he really not noticed the more than ninety injuries on Mousa's body? Had nothing inspired him to ask what had happened or whether the remaining detainees were being looked after properly? Had no alarms been sounded in his medically trained mind? Had he no duty to enquire into the welfare of these prisoners?

If the padre had been knee deep in morally dubious territory, Dr Keilloh was up to his neck. But he was released from the witness stand and allowed to return to his practice in Yorkshire. There was no space for his condemnation here at this court martial.

What, then, was there left of this trial?

20

IN THE NEW YEAR, THE prosecution case increasingly lacked conviction. An air of reprimand accompanied the introduction of the last witnesses brought against the defendants. The lawyers hadn't forsaken the case but a sort of sadness pervaded proceedings. It was as though the complexity and scale of the matter had become too burdensome after the amnesia and general closing of ranks. The army was not about to release its conscience to the court and the lawyers could do little about it.

By the end of January 2007 the prosecution had brought forward all the witnesses they could muster. They were faced by an inevitable challenge from all the defendants' representatives: they would each argue that there was no case to answer. The prosecution was flawed, they would say. Indeed, they would say more than that. There was outrage in the prepared speeches of defence counsel. Each of the QCs would use his eloquence in condemning the prosecuting authorities for ineptitude in selecting *their* clients to take the blame. It was a sham, a travesty of justice, they claimed. A 'bankrupt' case.

Tim Owen stood and addressed the charges his client faced. Manslaughter was the pre-eminent concern. He argued that Corporal Payne was entitled to use reasonable force against Baha Mousa in the moments prior to his collapse and death. Mousa was trying to escape. There was no evidence to suggest this was not true, and in the absence of such evidence there was every reason to accept that Payne believed he was trying to escape and could therefore be restrained. Owen said Payne had no training in the dangers of positional asphyxia and couldn't have been expected to realise the possible impact of his restraining methods: the knee in the back, the wrenching of the hoods about Mousa's head, the pinioning of arms behind his back. Payne couldn't have known the prisoner might die from his actions. He couldn't have been negligent, then, said Owen.

And with one of those matey asides which made the courtroom a place of fraternal familiarity for the lawyers and an alien and aggressive one for everyone else, Owen apologised for his hoarse voice.

'I'm afraid it's the result of over-exertion at the Emirates Stadium on Sunday afternoon.'

'I hope you were supporting the winner,' the judge replied.

'I was, My Lord.'

They resumed their serious dialogue without even the mention of 'football'. Perhaps that would have been too coarse.

Owen made the point that if the force applied was to include all assaults over the whole period of incarceration then others had to be in the frame for manslaughter as well, Private Cooper being one. Others had been involved both in the regime of ill-treatment and in the final restraint. At the least, they could be accused of aiding and abetting manslaughter, something which the prosecution had denied.

In any case, the medical evidence said that most of the injuries were survivable. That had been Dr Hill's position from the outset. Therefore the pressure applied by Payne with his knee to Baha Mousa's back must have been the act that caused death. At least that was the submission by Tim Owen. Even if this wasn't accepted, Owen argued that the violence in Payne's restraint was not as alleged. Aaron Cooper wasn't to be believed when he said Payne kicked Mousa in the head during those last moments. LCpl Redfearn backed up Payne's account. So too did Private Reader. Corporal Douglas never mentioned anything about a frenzied attack by Payne, a slap was all he said he saw. Against these accounts Cooper's evidence shouldn't hold sway. He was 'a self-confessed liar', Owen reminded the judge. Others too were known to have inflicted violence against the detainees, including most probably Baha Mousa. There had been evidence from all quarters of assaults at the hotel, where Payne had never gone, and back at Battle Group Main by a succession of soldiers. Corporal Payne could not in all conscience be landed with sole responsibility for the death. That would be wrong, Owen said. And even if the prosecution claimed that Payne ordered others to commit attacks, the testimonies from the Rodgers multiple were unanimous – none of them used more than a slap. So Payne couldn't be responsible by command either.

Mr Bevan attempted to repair some of the damage done by the conflicting and quite clearly untrustworthy evidence presented. He tried to emphasise the outrageousness of Payne's gratuitously violent conduct throughout Baha Mousa's detention. Despite all the lying and amnesia, despite the deliberate closing of ranks, the coordinated accounts, the pathetic failure to answer simple questions, despite the concerted efforts to justify the ill-treatment of prisoners, despite the obvious acceptance of dubious if not unlawful interrogation techniques throughout many quarters of the army, Bevan could still point to the violence that was undoubtedly done by Payne to the men in his charge. Would that be enough? Before these questions could be placed before the jury panel, the judge had to decide whether there was *any* case to answer. If he said 'no' then the accused would walk free there and then.

Arguments for all the other defendants had to be heard first, though. So, in fairly quick succession, the barristers for Crowcroft, Fallon and Stacey stood to deliver their harangues against the prosecution, the fatally flawed prosecution, as they would have it. Yes, the wounds inflicted on the detainees may have been proven, but the point in issue now was *when* they had been inflicted. Richard Ferguson QC said that no evidence had been adduced that could prove beyond reasonable doubt that his client had committed any *particular* injury on any *particular* detainee. The testimony of the medics showed that they had gone into the detention facility in the afternoon of the first day and they had spotted nothing untoward. As the only strong evidence of abuse appeared after the prisoners had arrived at Camp Bucca, then there was no independent evidence which could connect Crowcroft, Fallon and Stacey with those injuries. Private Jonathan Lee was the only person who said he had seen these three act violently against the detainees. Leaving aside that witness' credibility, had enough been shown that inhuman treatment, a war crime, had been committed? Ferguson argued that no reasonable jury could conclude that.

Mr Cox made similar submissions on behalf of Kingsman Fallon. He said the charge was wrong. He had asked before why it was that the fourteen members of A Company paraded as prosecution witnesses weren't charged as his client had been. And the answer offered was that Crowcroft and Fallon had used 'unreasonable and excessive force, at times gratuitous violence and showed no mercy upon those detainees'. In other words, Cox said, some force was deemed acceptable – the force used for stress positions, for instance – and that Fallon and Crowcroft had exceeded that. Where was the evidence for such an accusation? When had it been proven that the prisoners' treatment was unacceptable during the first eight hours of their confinement and acceptable thereafter? How could that square with the detainees' evidence that beatings had continued for the whole of their time at Battle Group Main? And if the prosecution's case was that the conditioning amounted to inhuman treatment, then surely all those who had carried it out, all those guards of A Company, had to be condemned too. The selection of Crowcroft and Fallon and not any soldier from the Rodgers multiple, who had been with the detainees first at the hotel and then from 6 pm on the first day until the morning of the third day, made no rational sense.

As if this were not enough, Cox could point to the evidence presented that Fallon and Crowcroft were not the only people in the facility during that first afternoon. Corporal Payne was there. Others came and went. If Private Lee's evidence was *not* to be believed, then there was nothing concrete to connect Crowcroft and Fallon to specific alleged assaults. If there was no direct evidence, the case had no substance. It would be brutally unfair to proceed.

Mr Baker on behalf of Sergeant Stacey, now accused of a single 'common' assault, made much the same points. He also had a direct comparison to make. LCpl Slicker had admitted in evidence that he had been punished by his regiment for assaulting a detainee and had escaped with a fine and no court martial. He had been the man arrested after being identified by Scott Hughes, he of the *GMTV* escort crew. So, why had Slicker been treated so leniently whilst Stacey had been subjected to the full intensity of prosecution? Given that Private Lee was the 'very unsatisfactory' witness of Stacey's misdemeanour, could a prosecution really be allowed to continue? It was surely against the interests of justice, claimed Mr Baker.

Those were the main arguments levelled by those accused of physical violence. The turn of those alleged to have acted in neglect of duty followed. These hinged on the testimony of Major Royce. Once 'conditioning' had been sanctioned as standard operating procedure, how could Major Peebles or the others be condemned? There was nothing to prove that Colonel Mendonça had knowledge of the mistreatment being meted out and the award of the DSO plainly suggested senior command saw nothing personally at fault in his handling of the detainee affair. And if it was accepted that Major Peebles had not been negligent then Colonel Mendonça couldn't

have been negligent. The whole case of neglect would fall like an elabo-
rately constructed house of cards.

The judge listened to the submissions and then retired. He gave himself
two weeks to decide whether the case would be allowed to continue.

2 1

TUESDAY 13 FEBRUARY 2007. MONTHS had passed in this acoustically
confined room with its blue carpet and soft-beech-coloured desks.
The smell of newness had been supplanted by the sweat of papers
and lawyers. Mr Justice McKinnon appeared, all in the chamber rising in
traditional respect, sat upon that leather upholstery and gazed about him
with serious intent. He was to deliver his judgment on the 'no case to
answer' submitted on behalf of all the defendants.

It was a difficult task. There was ample evidence that something seriously wrong had occurred at Battle Group Main in the middle of
September 2003. But that wasn't the crucial factor. McKinnon had to
ask himself whether, by that point, the prosecution had provided sufficient
evidence for a reasonable panel to convict. He had reviewed the charges
carefully and the testimony provided and assessed each defendant in
turn.

It was a moment of controlled climax. He considered the case against
Crowcroft, Fallon and Stacey. He said that the charge of inhuman treatment for the first two and the charge of common assault for Stacey
rested almost completely on the evidence of Private Lee. But, McKinnon
said, Private Lee was incapable of being believed; the unexplained
inconsistencies, the downright lies, the unreliability of the witness meant
his evidence had to be disregarded. His evidence was 'useless'. The
judge said that in the absence of any other testimony Crowcroft and
Fallon were in precisely the same position as all those other guards at
the detention facility, those members of A Company and all those men
and women who had wandered in and contributed to the violence
inflicted on the prisoners. He said there was no reliable evidence against
the two soldiers which could possibly allow a panel to convict them of
a war crime.

He considered the position of Colonel Mendonça. He said the prosecution's case was wholly constructed around the assumption that conditioning

of prisoners prior to interrogation had never been approved by Brigade. But, he said, the evidence of Major Royce, credible evidence, suggested that the regime had been sanctioned by intelligence and legal officers. He said Colonel Mendonça was entitled to rely on that advice so couldn't be said to have failed in the discharge of his duties. He said no panel could convict of negligently performing a duty on that ground.

And then he spoke about Corporal Payne and the charge of manslaughter. He said the prosecution had failed to prove that Payne had been responsible for Baha Mousa's injuries, in whole or in part. He said there were too many others who might have inflicted one or more of the ninety-three wounds found by Dr Hill. He said the prosecution had failed because they had argued the death was the result of an accumulation of harm, culminating in that last attack in the detention facility's middle room. He said the prosecution had failed because Payne was entitled to use force against Baha Mousa during those last moments: there was reasonable belief that the prisoner was attempting to escape. He said Payne was indeed obliged to use reasonable force in such circumstances and the prosecution had failed to prove that he had gone beyond what was reasonable. He said the act of escape had broken the link between any earlier violence and the aggressive restraining methods used. He said conceptually and evidentially the prosecution had failed. He said the charge of manslaughter had to be dismissed; no panel of jurors could properly convict on the basis of what had been heard.

It was here that the ever-so-strained link between the wrong that was perpetrated on that day in September 2003 and the proceedings before the court tore apart. The *death* of Baha Mousa had been the reason why investigators took notice, the reason why politicians had to be prepared to defend the armed forces, the reason why the press showed any interest, the reason why army commanders and lawyers displayed moral anxiety over the behaviour of British troops. Would anyone have noticed otherwise? A few bruises and spells in intensive care, perhaps a broken rib or two, stories of sexual humiliation easily denied; would those have provoked the slightest interest? For all the terror felt and harm suffered by the surviving detainees, would their accounts have attained more than passing recognition amongst the British or world's press?

Those rhetorical questions were redundant now. Even though the judge ruled that there was still a case to answer for Major Peebles and

WO Davies (the two in charge of interrogation), the thread attaching the trial to Baha Mousa was severed. All that remained was the tarnishing of a regiment and legal arguments about the legitimacy of conditioning.

22

COLONEL MENDONÇA MARCHED OUT INTO the tepid air of Bulford camp, emerging through the doors to the chamber like a bull. His arms were rigid by his side, his fists clenched, his shoulders slightly hunched as though he could barely prevent himself from charging at the gathering of reporters and photographers in vindication. He stood to attention and spoke.

It was 15 February 2007. The speech that Mendonça gave was short, controlled. He had served his country for twenty-five years, but the last two had been 'difficult' for him. He still believed his soldiers had done 'enormous good' in Basra. And he thanked those members of the services and the public who had supported him throughout the last few months. Now his family was his priority, who had 'borne the brunt' of the pressure of the trial. His wife, who had told her agonised story about the pressure on her and her husband nearly two years before, and his two young sons; they were waiting for him.

At no point during the colonel's short speech was there mention of Baha Mousa and *his* family nor of the two young boys who would never have the benefit of their father's return. The colonel could say no more because the court martial was continuing. Major Peebles and WO Davies, the two men closest to the imposition of conditioning on the detainees, hadn't had their cases dismissed by the judge. They still had a case to answer, although an aura of futility filtered through each submission and question from then on. In something of a twilight period the case continued without the panel hearing the judge's reasons for dismissing the charges against the other five defendants. Not that they would have been clueless about those reasons. They had heard enough mendacity, enough unbelievable claims of memory loss, enough officer corps cavilling about the legal framework for interrogation, to guess what had happened. Such was the fragility of evidence against the defendants that the panel must have been mystified why the tactical questioners remained on trial. Their duty was passive.

They had to be there to the end, whenever that was served up to them.

Nearly a month later the judge was able to sum up the case against the remaining defendants. At 10.35 am on 13 March 2007 the panel were asked to retire to consider their verdicts. They were gone four and a half hours. Seven months of evidence and argument, seven months of submersion in document mountains that would have tested the most ardent researcher, seven months of tedium and drama and banal legal dialogue were reduced to four and a half hours of deliberation. The panel returned to the chamber at 2.50 pm and pronounced Major Peebles and WO Davies 'not guilty'.

The two men were released. All that remained was for Corporal Payne to be sentenced for his admitted inhuman treatment. His plea in mitigation was accepted in part. The panel acknowledged that he had acted on orders and those orders necessitated the use of force. It didn't accept that he was guiltless. Payne may have been placed in an invidious position in being required to apply 'conditioning'. But they were satisfied that his violence had been gratuitous and excessive. No doubt the admitted story of 'the choir' broke any thought of force deemed necessary. The panel decided the reputation of the British Army had suffered as a result and he was sentenced to twelve months' imprisonment and discharge from the services.

The case was finished. A twelve-month sentence for one man. Everyone else free to return to duty or their civilian occupation. It was as if Battle Group Main hadn't existed, as though those days of beatings and kickings and humiliations in a sweat-drenched Basra had never occurred.

Part 4

EPILOGUE

A LITTLE OVER A YEAR *after the disastrous, often farcical, court martial, Des Browne, then Secretary of State for Defence, told Parliament that a public inquiry would be held into the death of Baha Mousa. Browne's decision wasn't an act of grace. In 2007, the House of Lords had handed down a judgment that the Human Rights Act should apply to the death in custody of Baha Mousa and the government was obliged to undertake a proper inquiry into the matter. The court martial and preceding Royal Military Police investigation were deemed insufficient.*

In making his announcement on 14 May 2008, Browne spoke of his determination to do everything he could 'to understand how it came to be that Mr Mousa lost his life'. The army, he said, had 'no wish to hide anything'. It had already looked critically at itself. And it had made changes. Brigadier Aitken had produced a report as instructed by General Sir Michael Jackson following the Camp Breadbasket trial fiasco in February 2005. It had taken the brigadier nearly three years to produce his thirty-six-page review. He felt confident that only a 'tiny' number of allegations were credible. He warned that 'it would be a mistake to make radical changes . . . unless there was clear evidence that the faults we were seeking to rectify were endemic'. According to him, they were not.

The full public inquiry called by the Defence Secretary would have to look again. On Monday 13 July 2009 Sir William Gage took his seat as chair of the Baha Mousa Inquiry in London and called upon Gerard Elias QC, counsel to the inquiry, to present his opening address. Three topics concerned him: how Baha Mousa died; the nature of training and command; and what lessons could be learnt from the whole affair. That was the territory he could cover. He wasn't allowed to consider the investigation and its quality. He wasn't allowed to consider the court martial and its failings.

Still, over the succeeding months, all those witnesses in the court martial and many more besides were called to take an oath and answer questions. There was again a bevy of barristers representing everyone from the Ministry of Defence, the army, the individual soldiers and the detainees, to test the evidence. But none of the lawyers would be allowed the licence to cross-examine with the ruthlessness that the court martial QCs had displayed. Public inquiries may be intrusive but they aren't trials. Their purpose is to find out the truth, not to prove a charge or 'break' a witness.

After more than two years of scrutiny, hundreds of witnesses called, all at a cost in excess of £12 million, the Gage Report was delivered in September 2011.

1

S IR WILLIAM GAGE ENTERS THE press room in Furnival Street, London at precisely 11.00 am Thursday 8 September 2011. He sits with his glasses pushed severely to the bridge of his nose. He is ready to introduce his report. This document, this *thing* that will release his judgment on the killing of Baha Mousa is before him, thick with print. Three volumes of meticulous and measured prose. There is not a word that hasn't been checked and rechecked for its appropriateness. Nothing must interfere with the credibility of the findings. It is a major piece of scholarship. Although he may not have been allowed to scrutinise the RMP's investigation or the adequacy of the Army Prosecution Authority's attempt to put men of 1QLR on trial, there are many other things he can say. He has heard hundreds of witnesses, some changing their stories, some still gripped by amnesia, some apologetic, some dissembling, some embarrassed, some brutally direct, some emotional. Two hundred and eleven people have given testimony under oath. Among them were the surviving detainees, invited to tell their stories once again. And Daoud Mousa was finally allowed to speak about his son and testify to the circumstances of his arrest. This time he and the detainees were treated with deference, even exaggerated respect. They were accorded honour as truth-tellers, as victims who were helping the inquiry rather than doubtful witnesses with dubious motives.

And now Sir William Gage is able to present some sense of what happened to Baha Mousa and the other men taken to that British Army base in downtown Basra that hot September day in 2003. Whether it is *the* truth, whatever that means, is perhaps irrelevant. It is as close as one can reasonably expect given the imperfection of memory and the complexity of individual motives. Despite the heaviness of the report, the thickness of its 1,000 and more pages, the account given by Gage feels weightless in comparison to the life that was taken and the bodies and minds affected. It deserves reiteration nonetheless as fact and no longer conjecture.

SATURDAY 13 SEPTEMBER 2003; THE 1ST BATTALION of the Queen's Lancashire Regiment prepares for a mission. It is called Operation Salerno. Intelligence has provided information that suggests hotels in Basra are used by criminal gangs and former regime loyalists (terrorists by another name) as safe houses. Several are identified across the city. One is called the Hotel al-Haitham. These hotels are to be searched. There is a list of suspects attached to the orders. It is to be a 'soft knock'; no forceful entry, merely a peaceable operation to see what can be found. The level of anticipated threat is low.

Call Sign 10A of Anzio Company in 1QLR under the immediate command of Lieutenant Craig Rodgers (the Rodgers multiple) is ordered to search the Hotel al-Haitham. They make their move at 6.30 am on Sunday 14 September, coinciding with soft knocks on other hotels in the area.

The entry to the hotel is quietly done. The troops come in through the front door. They are met by the hotel cleaner and part-time nightwatchman. He takes them to the reception where Baha Mousa is on duty. There is no resistance and no violence. Everyone is relaxed.

The nightwatchman shows members of the multiple the hotel rooms. Some doors need to be unlocked. All are searched but nothing is found. Lieutenant Rodgers discovers three guns at the reception desk, although this is not unusual. Hotels and businesses and private citizens keep weapons for protection. Criminal gangs can attack and there is an expectation that one will have to defend oneself. This is Basra.

But then the tranquil atmosphere suddenly alters. A locked toilet door behind the hotel manager's office is forced open because the key cannot be found. Inside, the soldiers discover a small collection of weapons: two grenades, a sub-machine gun and sniper goggles. These are beyond the necessity of self-defence. And then the soldiers realise that one of the co-owners of the hotel, who was in reception when the search started, has disappeared. A member of A Company has seen him walking out of a back door and away down an alley. It is not just suspicious. It is indistinguishable from admission.

The two events, the finding of weapons and the 'escape', provoke an intense reaction. Although the hotel guests have been the focus, now it is the employees. If the hotel's co-owner was worried enough to run away then who amongst the staff are also implicated?

Within a few minutes the soft knock has turned hard. Baha Mousa as receptionist is asked questions, but he cannot tell them anything. He does not know where the key to the safe is. The soldiers become irritated. All the hotel workers are rounded up and pushed on to the floor of the hotel lobby, face down. They are kept there for about an hour, maybe more. Whilst lying on the floor the detainees are nudged, kicked, perhaps stamped on. It is rough, though probably not severe.

At around 8 am Daoud Mousa arrives at the hotel to pick up his son. He walks up to one of the entrances and through it sees three or four British soldiers breaking into a safe. They reach into the safe and draw out packets of banknotes. Some they put in the side pockets of their uniforms or inside their shirts. Daoud Mousa sees this. He approaches a soldier guarding the hotel entrance. He asks to be let in as he wishes to report a crime. He is taken to Lieutenant Michael Crosbie, whom he tells of the thieving. Mousa sees one of the looting soldiers in the lobby and points him out to the lieutenant. Crosbie calls the soldier over and checks his pockets. A packet of banknotes is found. The lieutenant takes the notes from the soldier and orders him to get out of the hotel. The soldier's name is Private David Fearon ('Dinar Dave' he is called later by one of the others in his squad), a member of the Rodgers multiple.

Daoud Mousa sees his son lying on the floor amidst all the hotel employees. He asks the lieutenant to release Baha, but Lieutenant Crosbie cannot or will not do that. He gives Daoud Mousa his mobile phone number and says he can call him later. Daoud leaves the hotel and waits outside.

After lying on the lobby floor, the hotel employees are taken to the toilets behind the reception desk. Perhaps it is thought the men are too visible to the crowd gathering outside the hotel. They are conscious that the locals can become inflamed. A riot would jeopardise the operation so it is better to take the suspects out of sight.

The hotel employees are put in the toilet cubicles; later it will be claimed that they are forced to sit over the open holes, that the toilets are flushed so that water and excrement splash over them. Some will say they are punched as well. At least a couple of the hotel workers are questioned in the toilets; do they know where the escaped co-owner has gone? The soldiers discover the man's home address and a squad is sent to raid the property. But he is not there. He will never be apprehended. Another squad is sent to arrest another of the hotel's co-owners, the escapee's father, at his home. The restaurant manager accompanies them to show the way. The co-owner and his other, younger son are rousted from their beds and thrown into a truck.

At 9.50 am Major Englefield, who is the officer commanding the whole of Anzio Company, radios for instructions on how to move the people arrested in the hotel. Englefield notes that a TV camera crew is outside the front doors. He is told to use plasticuffs but no hoods. The logs record the decision to apply 'our normal methods bar sandbagging'. The guards are told to be more gentle than usual.

One by one the hotel staff members are taken to the army trucks waiting outside. Six (including Baha Mousa) are taken straight to Battle Group Main, the headquarters of 1QLR. Three, the hotel's co-owner, his eighteen-year-old son and the restaurant manager, are taken initially to Camp Stephen by Lieutenant Crosbie and a guard from members of the Hollender multiple of Anzio Company. These three are kept at Camp Stephen for about two hours.

During that time the eighteen-year-old is forced to do physical exercises in the sun. According to his testimony he is made to jump up and down until he collapses. Then he is pushed into a kneeling position in the gritted courtyard with his trouser legs pulled up so his knees are bare on the sharp little pebbles. Although his father can corroborate this treatment, no soldier will recognise the boy's account.

Eventually, the three Camp Stephen detainees are taken on to Battle Group Main to join the six other hotel employees.

When the initial six detainees, including Baha Mousa, are driven into the Battle Group Main compound at 10.40 am, they are taken out of the Bedford army truck and immediately marched into the detention block. Hoods are put over their heads, plasitcuffs are kept tight on their wrists and they are all subjected to intense verbal abuse: swearing, shouting. There are a large number of soldiers involved or watching. Private Slicker is there. So is Colour Sergeant Huxley, Private Felton, Corporal Stacey, Private Lee. Many are just interested to see who has been arrested. Many 'mill about' with no particular purpose.

'Conditioning' starts soon after the detainees are herded into one of the two rooms of the detention block. They are all forced into stress positions. They are made to sit against the walls, in 'ski' positions, legs bent at forty-five degrees, their cuffed hands stretching out in front of them. If they sink downwards, the guards yank them up again, by their collars, by their heads, all the while shouting insults and abuse at them. Corporal Payne unleashes a stream of invective: 'Get up, you fucking ape', and variations on that theme. There are punches and kicks too, according to the detainees. With the hoods over their heads they cannot see the blows coming. A second hood is added, perhaps to stop the detainees seeing their assailants through the hessian. The effect of the doubling is to increase the sense of suffocation, to disorient even more.

Once the general interest in the detainees dissipates and sightseers leave the detention facility, Corporal Donald Payne and privates Crowcroft and Fallon are left largely to themselves for the Sunday afternoon. Corporal Payne is in charge. The other two are assigned initial guard duty by Corporal Stacey, who stays in the detention block for a little while but then leaves to return to Camp Stephen.

At some point the 'choir' is devised. The detainees are lined up and struck in succession (some say tapped, some say hit) so as to elicit a cry, each man making a noise at a different pitch. It is an invention of Donald Payne. There are sustained and heavy assaults committed throughout the afternoon and the 'conditioning' is maintained. This is the 'softening' ordered prior to interrogation.

Several people wander into the detention block on that first afternoon. Lieutenant Crosbie is one. He is shown the choir but does nothing about it. He walks out of the detention block because he senses it is 'distasteful'. None of what he sees is sufficiently troubling for him to speak to anyone. He is an officer, young (twenty-four) and inexperienced.

Another officer, Lieutenant Douglas Ingram, Anzio Company's 'Watchkeeper', enters to collect the names of the detainees with the help of an interpreter. He sees them being 'conditioned', in their stress positions. When one detainee fails to answer a question, one of the guards (either Crowcroft or Fallon) punches him. Lieutenant Ingram yells at the guard. But he does nothing to stop the stressing. He will say later that he reports the punch to Major Michael Peebles, but that is the extent of his intervention. The conditioning is normal procedure as far as Peebles is concerned and he takes no action.

Others enter the block as well: SSgt Roberts, the commanding officer's bodyguard, who some see hit one or more of the detainees; CSgt Huxley is there at some stage too. And so is Major Peebles. He will admit to telling Fallon and Crowcroft early on that the detainees are suspected of involvement in the death of the three RMP officers a few weeks before. He will admit that the inflammatory remark, wholly unsubstantiated by any evidence, may have provoked some if not all of the violence.

Crowcroft and Fallon are on guard duty until about 7.00 pm. Then members of the Rodgers multiple arrive at the detention block to take over the guard. Fifteen to twenty soldiers pile in. Some of them will say later that the detainees are in a bad way already, roughed up. They are shown the 'choir'. Whatever the detainees' condition, the appearance of several of Rodgers' men leads to a sudden free-for-all abuse. According to the detainees, the new guards lay into them in a mass frenzy of punching and kicking.

For about thirty minutes the soldiers assembled in that little detention block cross some kind of line in their minds. Without any command restraint, without any moral restraint, the detainees become hooded targets for all to attack with impunity.

Privates Reader and Cooper, Lieutenant Rodgers, LCpl Redfearn, privates Aspinall, Appleby, Allibone, Kenny; they are probably all there. So too Private MacKenzie. Maybe others from the multiple enter the detention facility. Maybe all of them. But who assaults who remains obscure. The defenceless detainees endure the beatings from unknown boots and fists. They are hooded. They cannot see.

Once the frantic battering during the handover has subsided, the multiple settles down to guarding the detainees. Not that the violence ends. Conditioning is maintained.

The soldiers from Rodgers' multiple take it in turns to act as guards. They are split into 'stags'. At some point during this first evening both Lt Col Mendonça and Major Englefield visit the detention block. Neither will recall seeing anything to concern them.

At about 9.30 pm on this first night another detainee arrives for questioning. Ahmed Maitham is arrested by B Company in an incident wholly unrelated to the hotel raids. He is picked up when his car is pulled over and searched and three AK-47s are found in the boot. A guard takes him to Battle Group Main where he is thrown in with the other detainees. He is hooded and subjected to the same treatment as the others.

SSgt Davies is called from Brigade HQ to undertake questioning of all the detainees. He is supposed to finish his task within fourteen hours so the prisoners can be taken to Camp Bucca for more sustained interrogation. He arrives at Battle Group Main early on the Sunday morning and is briefed by Major Michael Peebles who is in charge of processing the prisoners. But despite the supposed urgent need for intelligence, Davies takes nearly twelve hours to call the first detainee.

The questioning follows the harsh shouting techniques learned by Davies at Chicksands, the military intelligence training centre back in Britain. But an added facet is the use of a powerful generator in the courtyard outside the block where the interrogation room is housed. The youngest of the detainees is deemed uncooperative and forced to kneel down in front of the generator as it blasts out hot air and drones incessantly and piercingly in his ears. The boy will claim he is hit and pushed up close to the machine. He is kept like this for an hour and a half. His soldier guard, drinking a mug of coffee or tea, laughs at him. This is all part of the 'naughty schoolboy' routine Major Peebles will say later is to encourage the boy to cooperate. Several people see the punishment being administered. No one intervenes.

SSgt Davies continues with questioning the detainees one after another long into the evening until Sergeant Smulski, another tactical questioner from Brigade, comes to help him. Smulski arrives about midnight on the first day. He takes over questioning duties from Davies after he has watched his colleague perform an interrogation session and has had a look at the detainees in the detention facility. Whilst there he instructs the guards to maintain the conditioning techniques he has also learned at Chicksands. He tells them to use the iron bar to bang against the tiled floor or walls. The noise will wake anyone.

Over those night hours the questioning continues and so too the abuse back in the detention block. The young boy appears to get the roughest treatment. Not only is he the brother of the man who escaped from the hotel, but he is also identified as the most likely to crack. Visibly frightened, he is pushed alone into the middle room, the old latrine, and kept awake. Soldiers smash the iron bar against the floor and tiles. The boy will say later he is urinated upon by one of the soldiers. He is kicked and punched repeatedly. It is all designed to weaken his resolve. Not that he has anything useful to say. He doesn't know where his brother has gone.

By the morning of the second day, Monday, the condition of the detention rooms is disgusting. Urine, faeces and sweat mingle to make the hot air putrid. The detainees are slumped on the floor, their clothes ripped, bruises and bloodied noses visible when their hoods are taken off. Some have soiled themselves.

Throughout the day their abuse continues. It is persistent. The 'choir' is performed with toddler-like repetition to show visitors. Many people witness it, many people see it acted out and do nothing, among them the crew from media operations, LCpl James Riley, LBdr Richard Betteridge and SAC Scott Hughes.

The padre, Father Peter Madden, drops in. He doesn't witness the choir but does see the squatting and the hoods. He will claim later that he sees little untoward with the stress positions or the sandbags over the detainees' heads. He does nothing to ensure these Muslim men are accorded any facilities for prayer.

Captain Seeds also probably comes in. Whenever it is, he sees the conditioning going on and thinks nothing of it either.

Lieutenant Rodgers is another caller. He is the commander of his multiple but exercises no restraint on his men in the treatment of the detainees. A number of others, non-commissioned officers, officers and men, visit. They cannot fail to notice the condition of the place and its prisoners. Some even contribute to the assaults. SSgt Roberts delivers a karate chop to one of the detainees. One or two are disturbed and report what they see to men more senior than them. No one makes a stand, though, and no one acts on information provided.

During this second day, attention turns to Baha Mousa. Perhaps he is targeted because of the report made by his father about the theft of money at the hotel. Some of the soldiers will later develop a story that he keeps trying to escape from his plasticuffs and remove his hood. But whatever the reason, in the afternoon Baha Mousa is removed from the others and placed in the middle room and ex-latrine, taking the place of the young boy.

Baha Mousa is heard screaming at various times after he has been taken away from the others.

Then at about 9 pm on the second day Corporal Payne returns to the detention facility to supervise a changing of the guard. Something triggers a frenzied attack by him on Baha Mousa. Payne will maintain that he comes in to find Baha Mousa trying to escape. He will say that he encounters him at the door of the latrine with his cuffs and hood off. Whether this is true or not, Payne's reaction is immediate and vicious. He pushes Baha Mousa to the floor. He puts his knee in the small of Mousa's back. He takes hold of his right arm and pulls it back hard. Private Cooper helps and grabs Mousa's left arm. There is a struggle, a frenetic attack by Payne. Payne will say Baha Mousa bangs his head. Cooper manages to get the plasticuffs back on. The restraint is violent and impairs Baha Mousa's breathing. He is lying face down on the ground, Payne's knee on his spine, his arms pulled behind his back.

There are several soldiers present. It is a confused and confusing episode.

Private Peter Bentham is there. He will tell a slightly different story from Payne. He sees Baha Mousa at the door of the latrine and pushes him back inside. It is only then that Payne appears and the two of them try to get his cuffs back on Mousa. Private Cooper arrives to take over from Bentham.

Private Reader is there. He sees Mousa receiving a good kicking. Payne and Cooper have a hold of the prisoner. After the struggle he tries to resuscitate him.

Corporal Douglas is there. He watches Payne punch the prisoner around the head. He watches, he will say, as Payne drags Baha Mousa about the room and flings him down.

LCpl Redfearn is there. He stands at the door and watches.

Private Cooper is there. He enters to see what the shouting is about. Cooper helps Payne in the restraint, attempting to hold on to Baha Mousa's legs. Mousa stops struggling and falls still although he is groaning feebly. Payne then lays into the prisoner, kicking and punching him repeatedly, about the ribs, about the head, forceful enough to propel Mousa's head against a wall. And according to Cooper, Payne grabs Mousa's head and bangs it against the wall a couple of times too. The attack lasts thirty seconds or so. Baha Mousa stops trying to protect himself.

Then Baha Mousa dies.

2

THIS IS THE ESSENCE OF Baha Mousa's killing. Even if embellished with the multiplicity of memories of the incident offered by numerous army and other witnesses, it remains a simple account. But it's a simple account nuanced with complex and resonating connotations. The death of this one man, however wrong within itself, isn't the totality of the story. Who killed him, how he was killed, are important questions which the inquiry was obliged to answer and within the bounds of reasonableness has done so. But judgment transcends these constraints. Why Baha Mousa was killed, what the killing signifies, are questions which are not so simply answered.

One sequence of facts is known, though. British forces operated in south-eastern Iraq for six years. Many thousands of Iraqis and 179 (136 in action) British troops were killed in the conflict. Many thousands of Iraqis were taken into custody. No one can be sure of the number of incidents when British troops abused Iraqis in their care. But it is certain that the killing of Baha Mousa and the abuse meted out to his fellow detainees was not an isolated incident. And it is doubtful that the numbers of occasions when abuse occurred were 'tiny' as Brigadier Aitken's report maintained. That word makes no sense. It's wholly dependent on context and perspective. Sir William Gage noted that 100,000 troops had served in Iraq and only a handful could have been involved in crimes such as the killing of Baha Mousa. Nonetheless many substantial allegations have been made since 2003. Phil Shiner's firm, Public Interest Lawyers, has alone dealt with at least 150 separate complainants involving multiple maltreatments. It is a depressing catalogue of abuse which they have uncovered:

Hooding
Stress techniques
Sensory bombardment
Sleep deprivation
Playing of loud music and DVD pornography
Electric shocks
Forced nakedness
Threats with weapons
Threats of violence to female family members
Death threats
Mock executions
Beatings with pistols, rifle butts, fists, helmets
Stamping
Forced kneeling on rough or hot tarmac surfaces
Masturbation by soldiers in front of detainees

Sexual humiliation by male and female soldiers
Deprivation of food and water
And killings

Just how many other incidents there were remains open to speculation but the evidence suggests several hundred, hardly a 'handful' or a 'tiny' number. Many may have been petty, thoughtless acts of violence, acts unrestrained by any common convention of respectful behaviour. Some were systematically abusive, like the treatment and interrogation of prisoners, euphemistically called 'conditioning'. All, though, have become indicative of a particular attitude which others will associate with the British Army if not Britain as a nation. This is the inconvenient truth which those who took the country to war rarely mention. It was not one that Sir William was able to consider.

3

DESPITE THE RESTRICTION TO HIS mandate, Sir William Gage could and did reach some conclusions about the whole system of detention and interrogation operating in Iraq. An entire section of his report was devoted to unpicking the army's protocol and regime. There had been, he said, a 'corporate failure' at the Ministry of Defence. When training manuals and prisoner-handling protocols had been drafted in the 1990s and after, they had forgotten the ban on those hooding and conditioning methods after the Northern Ireland experience in 1972. There was a wholesale failure to adopt a proper interrogation doctrine. No legal assessment had been made of the methods taught at Chicksands. When forces reached Iraq, they had no proper guidance on how to treat detainees lawfully and humanely, as they were required to do under the laws of war. Army lawyers stationed in Basra failed to appreciate those basic legal protections and ensure correct advice was given to the units who detained suspects. The implication was that a rotten system had developed. Rotten, that is, because no one cared sufficiently before or after Iraq was invaded to comply with those rules set down by the British government and by international laws fully endorsed by the British state.

Sir William made seventy-three recommendations in response to these findings. They almost matched the number of injuries so often quoted as suffered by Baha Mousa. The gist of them was that hooding and 'conditioning' should be banned; there should be no ambiguity in the manuals, in the training, in the standing orders.

Liam Fox, Secretary of State for Defence at the time of the report's delivery, immediately accepted seventy-two of the recommendations when he rose to speak in the House of Commons an hour after it had been published. The only one he baulked at was the banning of 'harshing' during interrogation. It was an odd one to single out for retention. The very word 'harshing' implies much more than anyone will ever define in the manuals. How many years will it take before someone misinterprets the rules again? How long will it be before the heat of the moment allows a chain of personnel to institute a new abusive regime?

Other than that, Gage avoided concluding anything rotten in the general state of the army or even 1QLR as a whole. He was certain the abuse of the detainees during 14 and 15 September 2003 was not an isolated incident, but nonetheless he made a number of negative findings: 'Although at times racist language was used by soldiers, there is no sufficient evidential basis to suggest that the violence was racially motivated'; 'the evidence does not demonstrate disciplinary failures so widespread as to be regarded as an entrenched culture of violence within 1QLR'. He was not entitled, nor did he have the resources, to examine any such culture across the forces serving in Iraq.

4

CAN ANYTHING ELSE BE SAID? Shouldn't the killing of Baha Mousa be put in context? Shouldn't the bravery and sacrifice of the troops sent to Iraq be recognised? Shouldn't the honesty and integrity of the majority be lauded *despite* the wrongdoing of a few miscreant soldiers? Shouldn't an attempt be made to understand the actions of the few instead of labelling them all and the army as a whole as criminal and abusive? Shouldn't a deeper type of truth be sought?

Almost all the objective judgments delivered on misconduct by British soldiers in Iraq (and every other conflict British troops have been engaged in since the Second World War) have been prefaced by positive answers to these questions. It must seem only reasonable to dismiss generalisations of collective wrongdoing before making specific and remediable condemnations. Sir

William Gage was no different. He stressed that the 'faults of some should not tarnish the image and reputation of the many'. But in adding this reservation there is a danger a deeper truth is obscured *not* sharpened.

It's more than arguable that the story of Baha Mousa and his abused compatriots, just as for others violated by British troops in Basra and elsewhere, is ineluctably bound to a pervasive condition of contempt, at best indifference. There is good cause to suggest the condition permeates the highest political office to the lowest military rank. There is good cause to suggest that it took diverse forms directed at both 'them' and 'us'.

For 'them' there was contempt or indifference for the application of rules, laws and even moral conventions – the ease by which methods of inhuman treatment were adopted and previous bans 'forgotten' is testament to this.

There was contempt or indifference for the people of occupied territories in the pursuit of political goals – the complacency shown towards planning for occupation after the military battle was won has been acknowledged within the army and government.

There was contempt or indifference for those purportedly being 'liberated', with wholesale suspicion and derision for the Iraqi people, frequently called 'Ali Babas', the 'Arabs', 'the insurgents'. Whether or not this inspired racist violence is a secondary issue.

There was contempt or indifference for unfamiliar lives and beliefs – the padre of 1QLR couldn't even recognise the religious needs of detainees and no one else demonstrated they cared either.

For 'us' there was contempt or indifference for the British soldiers, many just teenagers, little more than school leavers in most cases, sent in their thousands with limited consideration for their training and preparation in the conditions they would face, the behaviour they might encounter, the types of roles they would have to play.

There was contempt or indifference for instilling a clear moral framework to guide the troops during their time in Iraq – even the high command accepts that. Lord Dannatt, who was Chief of the General Staff in 2006, said there had been a lack of 'moral courage'. It wasn't evident at any level as a matter of course.

There was contempt or indifference for the capacity of military personnel to switch on demand from efficient killers to restrained police officers, peacekeepers, representatives of a nation supposedly intent on rebuilding a democratic, law-abiding society.

Across these variants on a theme, the contempt was both personal *and* institutional. Of course, individuals are culpable. That cannot be escaped. Corporal Payne may be the only person who has accepted that, even though only to an imperfect extent, but others made the abuse possible by actively engaging in violence or simply turning their backs on it. Nineteen soldiers of various ranks were identified by Sir William Gage as perpetrating some form of assault on the detainees. They came from across a range of units within 1QLR. And few if any (no officer and no ranking soldier, no medic and no clergyman) who saw what was happening in that dirty blockhouse has emerged from this story with much credit.* If not contemptuous of the plight of detained Iraqis they were indifferent, and if not indifferent, they lacked the courage or the will or the sense of duty to intervene seriously.

But was the ubiquity of contempt or indifference or lack of courage fostered by a more general and institutional condition? There was a command unwillingness (both military and political) to comprehend the consequences of going to war, of taking control of an alien country and its people. That much is common ground within the army, a story heard not just in the Baha Mousa Inquiry but also the hearings into the Iraq War as a whole overseen by Sir John Chilcot. There was a failure to prepare for the transition from battle to occupation. Although, in war, contempt may be a vital trait (it makes soldiers act with ferocity, it makes them kill when killing is needed), a proclamation of peace alters the moral coordinates. And perhaps there lies the paradox. The military *have* to be contemptuous. An aura of contempt for one's own safety and the humanity of others is a fundamental precept in the basic education of every fighting soldier and every commander sent to war. Why else, or how else, could one be prepared to kill? Respect for the enemy doesn't come into the equation, not when the soldier is immersed in an imagined world of kill or be killed. Any personification of the adversary would be ridiculous, counterproductive and contradictory to the soldier's purpose, something which surely can't be afforded in war. It's an issue of survival. Even if

---

* Sergeant Smith is perhaps the only one mentioned as someone who tried to question the continuing incarceration of the detainees and even ordered the hoods and cuffs to be removed. But even he does not escape censure, given that he could have done so much more to control his junior, Corporal Payne.

contempt isn't actively induced, it infuses every element of basic military training.

This should be expected. For all the laws of armed conflict, for all the cards handed to combatants with a printed list of their legal obligations on the back, savagery will still occur in war. No matter how sanitised the laser-guided weapons and precision bombing and carefully selected targeting techniques the army allows the public to see, the kill remains the desired end. To argue otherwise is simply foolish. There remains a 'bloodletting' in all wars, however the military and political machines try to present the violence as cleanly done. Contempt facilitates and makes feasible this martial culture. The enemy do not deserve to live because they are the enemy. That is the underlying rationality. To do harm to 'them' is to do right.

But what happens when soldiers are asked to suppress the ingrained training which dehumanises the enemy when in direct combat? What happens when soldiers are asked not to fight pitched battles against a known opponent, but act as policemen, builders, goodwill ambassadors, counter-terrorist troops? Can they truly suspend their embedded contempt?

One of the persistent stories surrounding the detainees at the time was that they were somehow responsible for the killing of Captain Dai Jones or the three RMP officers. It was a rumour generated by soldiers in Battle Group Main without foundation. Many people would later say that it gave them reason to wander over to the detention facility, look through the open doorways, walk inside and watch the ill-treatment perpetrated by the guards or even lend a hand themselves.

This stimulus, however ill-informed, can't explain the way in which the whole conditioning process and techniques used were so easily accepted and imposed long before anyone from 1QLR had been killed by insurgents. That there should be no one, no soldier, no officer, no clergyman, no medic, no RMP officer, no provost member, no one (other than one or two concerned lawyers) who would say *anything* against the ill-treatment or the regime of interrogation throughout Iraq at the time, suggests that contempt or at best indifference was deeply set. It overrode even basic moral judgment. And it wasn't confined to 1QLR. The Camp Breadbasket case, where 'Ali Babas' were subjected to humiliation and abuse by soldiers who believed they had been ordered to act in this way and felt sufficiently protected to take photos of their 'fun'; the 'wetting' of Ahmed Kareem and others, thrown into the stinking waterways of Basra as punishment for suspected thieving; the cases that have prompted two further public inquiries: these all point to more of a culture of contempt than any judge reviewing single abuses (such as in Baha Mousa's case) could confidently declare.

Such a culture was supported by the ineffective system of investigation in place. It was indifferent to the demands of a full, prompt and proper inquiry. No one would take responsibility for pursuing the culprits in an efficient manner. At every stage mistakes were made and a full and proper investigation failed to materialise. From the moment that the Special Investigations Branch of the RMP had notice of a death, the approach was slow and often flawed. No immediate dispatch of investigators to the crime scene. No scene-of-crime examination until far too late. No interview of key participants at the earliest possible stage. No identification or pursuit of suspects until weeks had passed, despite the fact that information about the men concerned was available from minutes after Baha Mousa was declared dead. No proper scrutiny of the senior officers involved. As Private Garry Reader testified, only the lowest ranks were expected to take the blame. 'Shit runs downhill' he was told; whatever happened, it would be the squaddies who would be held responsible.

Yet it must have been obvious to any investigator that errors of judgment, errors of professionalism, errors of decision-making had been displayed by many of the officer corps. The regimental doctor, Dr Derek Keilloh, who saw nothing. The hospital doctor who was willing to sign off a death certificate without seeing a body. The unconcerned regimental padre. The lawyers who couldn't agree on anything to do with the practices operated. The interrogators who thought they had been trained to condition detainees using techniques long banned by the armed forces. The line officers who couldn't bring themselves to intervene, who couldn't or wouldn't speak out against the abuse, who even saw it as acceptable. None of these men were pursued to good effect. None were punished apart from Donald Payne.

And when a trial of some sort did occur, what could the criminal justice institution achieve? Only seven men were brought before the court martial. Gage's report belied the adequacy of that decision. Nineteen soldiers were found by him to have committed one assault or another and numerous other officers and men identified as in breach of their duties. Much of the information Gage relied upon was available to the prosecutors in 2006.

As for the court martial itself further scorn was poured on the detainees brought to give evidence. In the interests of fairness the victims were either excluded from the process or subjected to the intensity of legal derision. Professional lawyers used every stratagem to decry their testimonies. The experience was deeply abusive. The institution was impotent in either approaching the truth or bringing those responsible to account.

5

L ORD DANNATT, CHIEF OF THE General Staff from 2006 until 2009, addressed the Royal Services Institute in February 2011 and lamented the moral vacuum in the modern British soldier. They need a 'moral education', he said, the 'core values' of 'courage, integrity, respect, loyalty, discipline and selfless commitment' had to be instilled afresh in recruits. Many of them came from 'chaotic backgrounds', he said, and hadn't been exposed to these values. Without this 'moral baseline', they may bully or abuse. In short, they couldn't tell right from wrong. That's what happened in Iraq, he said.

It was true that most of the rank-and-file soldiers in A Company of 1QLR, and no doubt soldiers in every other battalion, were not highly educated. They hadn't experienced privileged upbringings. For some, army life was an escape, a way of making something of themselves, a chance to avoid the banality of benefits, YOP training schemes, social housing. Some were undoubtedly immature (the average age of the Rodgers multiple was about twenty-one) but there's no proof that they were afflicted by chaotic backgrounds. Their actions both in Iraq and in the court martial were reminiscent of callow teenagers capable of acts of imitative cruelty. By all accounts this didn't stop them being forged into a good fighting unit. But they were never given the opportunity to fight battles against troop formations. In all their time in Iraq no uniformed opponents showed themselves. They had missed the invasion. There were no tanks to destroy. All those techniques of modern warfare that are often the explicit attraction in the recruitment advertising were redundant. The insurgents weren't being fought with high-tech equipment and sophisticated communications and diverse weapons 'platforms'. The battle, such as it was, was raw and so were they.

Of course, A Company 1QLR wasn't made up entirely of teenagers. Some soldiers had been on active service in other conflict areas, in Kosovo and Northern Ireland. These men were older, a little wiser perhaps and better able to adjust. Were their moral antennae any more developed than the rest of the men? Several non-commissioned officers showed no restraint, no moral sense, either.

And what about the officers? They were happy to go along with abusive practices of hooding and conditioning as part of interrogation. Some had reservations but none had the courage to raise an effective warning. The failure of values that Lord Dannatt alleged was just as much a matter for the officer corps as it was for the other ranks. It was particularly marked for those with additional professional responsibilities, like Dr Keilloh and Father Madden. Did Dannatt think they too were victims of 'chaotic backgrounds'?

Perhaps it's unjust to highlight the instances of illegal violence when contemplating the role of the military in Iraq. It infers that all members of the armed forces are guilty, if only by association, which is patently an absurd accusation. But this isn't the point. However 'good' the majority, however small the minority of wrongdoers, it isn't the condemnation of all for the suffering induced by the acts of the few which is the pressing moral issue. It's the institutional knowledge that these acts are inevitable in conflicts where soldiers are trained to be contemptuous of an 'enemy' but asked to somehow reshape that trained instinct, to act not as killers first and foremost, but as police, *without* any transition or preparation. It's this deep irresponsibility which begs for recognition and reform. Condemnation must surely be directed towards those in government and the armed forces who send troops to locations knowing that atrocities at some level will occur. It is predictable to the point of certainty.

Instead of recognising the inevitability, the official language is now of 'values', 'order' and commitment to standards. Instead of addressing the unavoidable, the design is to ignore it and then apologise for it later. Knowledge, act, apology is the institutional framework. It's a governmental culture that the British system has displayed for decades in its numerous dirty little end-of-empire wars. A deep-set contempt for others that infected military operations, that can be evidenced by the well-documented mass executions and floggings and torture in the Mau Mau revolt in Kenya; the deliberate targeting of civilian populations in Aden; the internment and shoot-to-kill policies in Northern Ireland; the torture in Cyprus; the killings, beatings and callousness in Palestine, Borneo,

Dhofar, Malaya, Oman, the Falklands. Not every participant was or is guilty. But who could argue convincingly that 'Britain' wasn't responsible for these wrongdoings?

6

WHATEVER MORAL AND POLITICAL QUESTIONS are left hanging after the inquiry, individual soldiers connected to the killing of Baha Mousa are unlikely to escape its consequences. The affair has dragged on for years and for some it hasn't finished yet. Their lives may have continued along normal paths, but escape has been difficult. Sir William Gage spared few in his designation of individual responsibility, although his criticism was limited to members of 1QLR, thus neatly corralling the moral contagion. How deep did his condemnations cut?

COLONEL MENDONÇA MAY HAVE THOUGHT THAT HIS troubles were over when he strode out of the court martial, exonerated in his and many others' eyes. But soon after the trial collapsed the army and the Ministry of Defence decided to resurrect the matter. An internal inquiry was ordered to look into the way in which 1QLR had been commanded. Mendonça remained under suspicion, liable to 'administrative punishment' as it was sweetly named. The news shattered any exhilaration Mendonça felt when the charges against him were dismissed. The thick cloak of suspicion was settling again around his army career. He resigned his commission and turned to the press to tell his story. The *Daily Mail* ran a sequence of articles in June 2007 reporting on the sense of betrayal and condemning the political interference in his case. They published photographs of Mendonça and his wife, posing as though for a fashion shoot, but with joyless faces. The quotes were replete with virtue: 'courage', 'honour', 'empathy', 'love'. It was a picture of an exemplary man and loyal family. But despite the incipient anger, it was the end of his devotion to the army. He couldn't escape the doubts over his integrity.

After leaving the forces Mendonça was (and appears still to be) promoted on the circuit of motivational speakers, advertising his abilities to galvanise men to work to the optimum. There is no shyness in referring to his appearance before a court martial. It's held out almost as a badge of honour; the dismissal of the case against him lauded as a vindication of his value. More recently he's turned to commercial management, acting as project director for Network Rail, assisting in delivering the new Crossrail line from Paddington to Maidenhead. A sedate employment. The absence of the excitement he experienced in Basra must be profound.

When Mendonça appeared before the inquiry he presented himself with unsurprising confidence. For Sir William Gage he was an impressive witness. No blame was directed at him for the infliction of particular violence. But he was condemned for a sequence of failures: he ought to have recognised the serious risk of inhuman treatment which hooding and stress positions in very high temperatures would have posed; he ought to have applied his mind to the whole matter of prisoner handling more acutely; he ought to have known what was happening in that detention block; he ought to have visited the detainees, discovered why they were still being held despite the rule that they should be transferred to Camp Bucca within fourteen hours of their arrest; and when he found out about Baha Mousa's death, he ought to have checked on the state of the remaining detainees. He was the commanding officer and his responsibilities couldn't be so easily delegated when it came to the welfare of those in his charge. Gage's findings were in direct conflict with the decision of the court martial. But Mendonça remains unpunished in law.

MAJOR CHRISTOPHER SÜSS-FRANCKSEN, Mendonça's second in command and the man who entered the detention centre with Major Seeds not long after Baha Mousa had died but who claimed he saw nothing particularly disturbing, became a manager at a Hewlett-Packard-related company in Reading. At the inquiry he said that he didn't think the detainees were obviously in serious distress when he visited the detention block. Gage didn't believe him. Süss-Francksen failed to discharge his duty that night, Gage found. He should have ensured those clearly suffering received medical attention and their welfare was maintained whilst they were imprisoned at Battle Group Main. Süss-Francksen has never answered for these failures.

MAJOR MICHAEL PEEBLES, THE MAN RESPONSIBLE FOR the system of prisoner handling and interrogation, did not resign his commission after

Iraq. He remained in the services and became a member of the Intelligence Corps, taking up a post in Afghanistan. Perhaps more than any other officer he was subjected to sustained critique by Gage. He found that Peebles ought to have known that the detainees were endangered by the system of conditioning he oversaw. It was an unacceptable failure. Peebles must also have known that the detainees were being subjected to serious assaults. That placed a heavy responsibility for all that occurred on Peebles' shoulders. His was a failure of duty that persisted throughout the period when the detainees were held at Battle Group Main, Gage determined. Again this contradicted the verdict of the court martial. Shortly after the Gage Report was delivered to Parliament, it was reported that Major Peebles had been suspended. But he remains unpunished.

The officer in command of A Company, Major Robert Englefield is also still in the armed forces. He has been promoted to lieutenant colonel. At the time, he was the man in charge of Lieutenant Rodgers and his multiple. He visited the detention block on at least a couple of occasions, although denied entering the building. He saw and heard nothing, or so he claimed. Gage called him an 'unsatisfactory witness'. But there was no evidence to prove that Englefield knew of the detainees' distress and Gage believed he could not be criticised for failing to look to their welfare, given that this was clearly the responsibility of Major Peebles. That doesn't mean he acted well in this affair.

Major Mark Moutarde was the adjutant for 1QLR. He couldn't remember whether he visited the detention block after he heard that a detainee had died. At the inquiry he conceded that it was highly likely he had gone there. If he had, and Sir William Gage found that he probably did, then he had a duty to investigate the state of the detainees and ensure their welfare was preserved from that moment. He did neither. Mark Moutarde's Facebook entry has listed Don Payne, Jorge Mendonça and Craig Slicker all as 'friends'. It means little perhaps, except that the links forged through service in distant wars are unlikely to be broken by scandal.

Lieutenant Craig Rodgers has also left the army. He has been pursued by the BBC TV programme *Panorama* but has stayed out of view on the whole, moving to Spain. He might have considered himself extremely fortunate not to have been prosecuted at the court martial along with Colonel Mendonça and the others. His command of the multiple at the centre of the arrest and guard of the detainees placed him close to all the abuse that was suffered. He has claimed, though, that he never saw anything untoward happening on the Sunday evening, on the Monday or even on the Tuesday morning. Gage found this unbelievable. He found that Rodgers knew much of what was happening, as it was happening. He found that he failed to report what was taking place, failed to intervene, and failed to exercise any proper control over his men. Gage said he had to accept serious responsibility for the ill-discipline. Rodgers has never answered for these failures.

FATHER PETER MADDEN LEFT THE ARMY AND has returned to work as a Catholic parish priest. He served the parish of St Mary's in Wednesbury and then moved to St Mary Immaculate in Warwick. He doesn't like to speak about the Baha Mousa affair. He's out of the army now, he says, as though a line of moral responsibility has been drawn for him.* Gage hasn't allowed that comfortable position to linger. He found Madden to be a 'poor witness' despite the fact that he had given his evidence under oath. He found that he must have seen the shocking conditions of the detainees during his visit. He found that Madden should have intervened or reported the matter to those in command. He 'did not have the courage to do either', Gage said. Newspaper reports stated that the priest would be spoken to by his archbishop. Whether he answers for his lack of virtue might be a matter for conscience rather than law.

THE HIPPOCRATIC OATH MAY BE AN ANCIENT, and some might say outdated, ethical framework for modern day physicians. But its call for doctors to keep their charges safe from harm and injustice remains as directly relevant as ever. Derek Keilloh left the army as a major. He's still a GP in Northallerton. His status as the senior medical officer in 1QLR placed him in a peculiar position as soon as he became aware

* I spoke to Father Peter Madden by telephone on Thursday 16 June 2011 but he said he wasn't interested in speaking about the Baha Mousa matter. He said, 'I'm out of the army now.' He refused to say anything more.

that there were detainees in Battle Group Main on Sunday 14 September 2003. He may not have known of the extent of their hooding and conditioning or the abuse they were receiving in the detention block. For Gage, he knew that the detention block was a squalid confinement which warranted careful medical concern. That procedures for looking after the health needs of all detainees were inadequate had to be partly (but not only) the fault of Keilloh. His inability to introduce a system that protected those under his charge, if only temporarily, was lamentable.

More profoundly disturbing was his refusal to admit that he saw any injury to Baha Mousa during or immediately after his attempts at resuscitation that would have alerted him to unjust harm having been committed. Other medics in that aid post saw the wounds. He said he did not. Nor did he see the injuries to the two other detainees who presented to him soon after the death. He prescribed minor pain relief and anti-inflammatory drugs as though that would suffice. It was a horrendous dereliction of his army duty and his duties as a doctor.

His failure didn't end there. Gage believed that Keilloh should have visited the detention facility to check on the health of the prisoners even before Baha Mousa's death. He knew they were there and could have guessed at the conditions. Immediately after the failed resuscitation he should have made 'the short walk from the RAP', as Gage noted with stiletto acuity, to see for himself how the remaining detainees were faring. And he should have reported the injuries he must have seen to those higher up the command chain. He had the opportunity. He had good cause. Keilloh now faces a thirty-day hearing before the General Medical Council's Fitness to Practice panel to answer charges relating to his actions in Battle Group Main all those years ago.

These men were the officers criticised by Sir William Gage. A couple of others, WO Mark Davies and Sergeant Smulski, were the tactical questioners on duty at Battle Group Main. Gage found that Smulski was 'singularly unprepared' for this task. Both men ordered forms of abuse: Davies by making the eighteen-year-old boy kneel close to the generator pumping out hot air into his face; Smulski by walking into the detention centre and reinforcing the harsh conditioning, including the use of the metal bar to keep the detainees awake throughout that first night. They knew what was happening and did nothing to prevent it. If anything, they gave the treatment an official air. They were the questioners from Brigade, the men who were supposed to know what was allowed and how far the guard could go. They may not have been aware of the kicking and punching but the rest of the abuse was fine by them. They believed that was what their training demanded.

Of course, all these men were not the ones positively identified as assaulting the detainees. Eighteen lower-ranked soldiers were listed as causing actual harm. Their names are familiar.* They appeared at the court martial. Whether they will ever answer for their willingness to attack the detainees, to watch and follow others, remains to be decided.

There is one more culprit to mention: the man Gage described as bearing a very heavy responsibility for the events in the detention facility which he oversaw. He was a bully, feared by the junior soldiers. He perpetrated much of the violence done to the detainees. He set the example for the others to follow. He initiated conditioning and maintained it with violence, instructing others to do the same. That violence was gratuitous, sustained and malicious. The choir was his invention, its repetition the act of a man who enjoyed inflicting pain. And at the end, his assault on Baha Mousa was intense and uncontrolled. It was a contributory cause to the death. There was nothing accidental about it.

That Donald Payne escaped conviction for manslaughter was probably more a matter of his lawyers' collective skill than the justness of his defence. He lied about his actions then and although he revealed much more before the inquiry, he was the primary perpetrator in Baha Mousa's death. After all that was learned at the inquiry, all the evidence of the multiple assaults, the choir, the enforced harsh conditions, the final uncompromising attack, a charge of murder or manslaughter no longer seems as ridiculous as the court martial concluded.

But is Donald Payne a scapegoat? Perhaps a distinction must be made. For all the ills and all the contempt evident across 1QLR and the army

---

* The eighteen are: Private Allibone; Private Appleby; Private Aspinall; Private Bentham; Private Cooper; LCpl Crowcroft; Corporal Douglas; Kingsman Fallon; LCpl Graham; Private Lee; CSgt Livesey; Private MacKenzie; Private Reader; LCpl Redfearn; SSgt Roberts; Private Slicker; Sergeant Stacey; and Private Stirland.

and the government in its conceited invasion and occupation of Iraq and treatment of that country's inhabitants, of course he is. For Baha Mousa's death, no.

7

F IVE DAYS AFTER Sir William Gage released his report, Baha Mousa's father, Daoud Mousa, appeared at a memorial lecture for his son in London. He shared a platform with Phil Shiner, his lawyer, and a couple of notable legal practitioners. The picture of Baha with wife and children was projected on to a screen behind his head. It was replaced later in the evening by photographs of Baha Mousa's post-mortem. Daoud Mousa, a slightly older but recognisably similar figure to his son (the same short, powerful, stocky build, thin moustache, tightly wired oil-black hair), turned around in his seat. He looked up at the vast images behind him and gazed at those large patches of bruising across ribcage and back, the vicious swathe of scraped skin about his wrists where the plasticuffs had bit and chafed, the face bloated with contusions. His son lay there, projected for the audience to understand the human reality behind the killing, the killing that provoked eight years of legal action for recognition that a man had been murdered and his most basic rights violated; the killing that forced a military police investigation and a largely fruitless court martial of seven suspects; the killing that compelled an internal army review, a full public judicial inquiry, the judgment of every court of significance in the British judicial system, House of Commons questions and reports, press articles, commentary and television documentary features.

Daoud Mousa crooked his neck and stared at the pictures for a long time.

And then he stood up, turning his back on the screen.

'*Asalam alaikum*,' he said. 'Peace be upon you.'

NOTE ON SOURCES

The case of Baha Mousa eventually became one of the most openly reported in British judicial history. Sir William Gage's inquiry examined thousands of documents and hundreds of witnesses. There was little hidden from view; transcripts of the hearings and statements of all those called before it were made public. Certain information was restricted, however, to protect the victims who did not want to be identified, or army witnesses who wished to remain anonymous in the interests of security. These strictures have been respected in this book for the same reasons.

ACKNOWLEDGMENTS

I would not have completed this book were it not for the help and encouragement of several people. Foremost amongst those is Phil Shiner, an inspirational figure who has continually challenged convention and indifference through his legal practice in the pursuit of justice. If the law is ever to gain a good name, it will be through committed and professional lawyers such as he.

Others have provided much needed assistance and reassurance in the production of the final text. By reading and commenting on its style and content (or giving me the heart to finish), Danny Friedman, Anita Mason, Maureen Freely, Patrick Bradley, Andrew Gordon, Laurence Williams, Norman Williams, Dan Franklin and Kay Peddle provided essential support, enthusiasm and advice.

But as ever, nothing would have been written by me had it not been for Kathy, Antonia and Claudia.